# Phosphorus in Soils and Plants

*Edited by Naser A. Anjum, Asim Masood, Shahid Umar and Nafees A. Khan*

Published in London, United Kingdom

Phosphorus in Soils and Plants
http://dx.doi.org/10.5772/intechopen.107701
Edited by Naser A. Anjum, Asim Masood, Shahid Umar and Nafees A. Khan

Contributors
Alex Odoom, Wilfred Ofosu, Anurag Yadav, Kusum Yadav, Eisa Ebrahimi, Mohammad Reza Ojani, Naser A. Anjum, Tahsina Sharmin Hoque, Deepranjan Sarkar, Rahul Datta, Mohammad Golam Kibria, Rafi Ullah, Nazeer Ahmed, Mohammad Anwar Hossain, Asim Masood, Krishanthi Jayarukshi Kumari Premarathne, Asanka Udaya Aberathna, Dilan Amila Satharasinghe, Anura Prasantha Jayasooriya, H. Rasika Niranjana Jinadasa, Sivabalan Manopriya, B.P. Ajith Jayaweera, C. Annesly Nandana Fernando, W.A. Dammika Vijitha Weerathilake, G. A. Prathapasinghe, Janitha Abeywickrema Liyanage, Valentin Nenov, Hyusein Yemendzhiev, Gergana Peeva, Shahid Umar, Nafees A. Khan



First published in London, United Kingdom, 2024 by IntechOpen
IntechOpen is the global imprint of INTECHOPEN LIMITED, registered in England and Wales, registration number: 11086078, 5 Princes Gate Court, London, SW7 2QJ, United Kingdom

British Library Cataloguing-in-Publication Data
A catalogue record for this book is available from the British Library

Additional hard and PDF copies can be obtained from orders@intechopen.com

Phosphorus in Soils and Plants
Edited by Naser A. Anjum, Asim Masood, Shahid Umar and Nafees A. Khan
p. cm.
Print ISBN 978-1-83769-035-0
Online ISBN 978-1-83769-036-7
eBook (PDF) ISBN 978-1-83769-037-4

For EU product safety concerns:
IN TECH d.o.o., Prolaz Marije Krucifikse Kozulić 3, 51000 Rijeka, Croatia,
info@intechopen.com or visit our website at intechopen.com.

# Meet the editors

Dr. Naser A. Anjum obtained a Ph.D. in Botany from Jamia Hamdard, New Delhi, India, and received post-doctoral research training at the University of Aveiro, Portugal; Agricultural Biotechnology Research Centre, Academia Sinica, Taiwan; and Aligarh Muslim University, Aligarh, India. Dr. Anjum has significantly contributed to the biology of plant/crop-environment adaptation. He has received prestigious research awards from reputed funding agencies, including the Portuguese Foundation for Science and Technology (FCT); the Council of Scientific and Industrial Research (CSIR), New Delhi, India; and the Department of Biotechnology (DBT), New Delhi, India. Dr. Anjum has published over 125 peer-reviewed papers, edited 18 books, and journal special issues. He is an elected fellow of the Linnean Society, London; Indian Botanical Society; International Society of Environmental Botanists, India; and National Academy of Biological Sciences, India.

Dr. Asim Masood has unveiled major insights into abiotic stress (metals, salt, and temperature) impacts and underlying physiological/ biochemical mechanisms in crop plants. He received a Start-Up Career Grant from the University Grants Commission, New Delhi, India, and DST-SERB Young Scientist Project from the Department of Science and Technology, New Delhi, India. His research interests include the involvement of ethylene in sulfur-mediated alleviation of cadmium, ethylene-mediated regulation of photosynthesis as a result of reduced glucose sensitivity, ethylene-mediated modulation of gibberellic acid-induced sulfur-assimilation under cadmium stress, and ethylene-mediated reversal of salinity impacts involving ethylene-abscisic acid coordination. Dr. Masood has more than sixty research papers to his credit.

Prof. Shahid Umar has identified the major strategies for developing abiotic stress-tolerant crops through modulation of tolerance mechanisms at both physiological and molecular levels, deciphered the importance of nitrogen metabolism in cereals and leafy vegetables, and enhanced nitrogen-use efficiency as well as developed a synergism between primary nutrients and growth under varied environmental conditions. Prof. Umar has nine edited volumes to his credit and has participated in research projects with funding from agencies, including the International Potash Research Institute, Switzerland. For his significant contributions to potassium nutrition management for the improvement of crop yield and quality, Prof. Umar has received several awards including the 2003 IPI-FAI International Award from the International Potash Institute, Basel, Switzerland, and the Fertilizer Association of India.

Prof. Nafees A. Khan has contributed immensely to abiotic tolerance mechanisms with a positive shift in the source–sink relationship. Mechanism mainly involved ethylene and its coordination with other major phytohormones-mediated induction of signal transduction and eventual modulation of the ascorbate–glutathione system. His path-breaking R&D contributions in the physio-biochemistry of abiotic stress tolerance strategies led to the publication of several journal articles. Prof. Khan was deemed a highly cited researcher for four consecutive years (2019–2022) and received the Clarivate India Research Excellence Citation Award in 2023. He has edited twenty-one books and served as editor/guest editor of leading publishing platforms. Prof. Khan is an elected fellow of the National Academy of Sciences, India (NASI), the Linnean Society, the Indian Society for Plant Physiology, and the Indian Botanical Society.

# Contents

# Preface

The burgeoning world population largely depends on plants and/or on plant-associated food resources. Plants need several nutrient elements at optimal levels to maintain their growth, metabolism, and productivity, as well as to sustain under adverse conditions. Phosphorus (P) stands second to nitrogen in terms of its essentiality as a plant macronutrient, as well as due to its involvement in all plant developmental stages, primary and secondary plant metabolisms, maintenance of membrane structures, and in the structural skeleton of biomolecules (including adenosine triphosphate (ATP), nicotinamide adenine dinucleotide phosphate (NADPH), nucleic acids, phospholipids, and sugar phosphates). Interestingly, under an inevitable wide range of abiotic stresses, P supply maintains cell turgidity and cell membrane stability, modulates plant–water relation, improves photosynthetic functions, enhances energy-rich compounds (including ATP, cytidine triphosphate (CTP), guanosine triphosphate (GTP), uridine triphosphate (UTP), and phosphoenolpyruvate (PEP)), and combats oxidative stress-caused consequences via maintaining the major enzymatic and non-enzymatic components' antioxidant defense system.

Though the concentration of P in soil is 2000-fold higher than that in the plant, most soil-P is unavailable for uptake by plants because most soil-P is fixed in the form of Al/Fe or Ca/Mg phosphates. Additionally, less than 0.1% of the total P exists in inorganically (Pi) available forms for plant uptake due to low solubility of Pi, slow diffusion, and high soil reactivity. Hence, most plants face the problem of P-deficiency in agricultural fields worldwide. To achieve the aforementioned benefits of P in plants, P-containing fertilizers are being added to agricultural lands to sustain high yields. However, only the least amount of the applied phosphate (20%–30%) is used by most cultivated plants, and the rest is lost, eventually causing water eutrophication. Moreover, mined rock P is the prime source of P used in agriculture worldwide, which in turn is adding varied toxic elements to the soils.

Given the preceding, this book, Phosphorus in Soils and Plants, provides a comprehensive overview of P in soils and plants, P dynamics in the soil–water–sediment environment, the major roles of P in the photosynthetic dark phase-biochemical pathways, major approaches for the sustainable management of P in agriculture to maintain optimum plant/crop growth and safeguard the environment, main mechanisms underlying the role of P in the regulation of plant–microbe interactions in the rhizosphere for optimum plant growth and metabolism, literature on the role of microbial phosphate-solubilization in management of soil and plant nutrients, and insights into P-recovery through waste transformation to implicate the research outcomes for an alternative fertilizer.

We believe that this volume will serve as an important resource for plant biologists involved in teaching or research, as it provides major insights into P in soils and plants.

We take this opportunity to express our thanks to IntechOpen for providing a platform for showcasing our work. We are also thankful to the Author Service Manager Mrs. Karla Skuliber for her efficiency and guidance, which helped us to accomplish this book project. Our sincere gratitude also goes to the chapter authors who contributed their time and expertise to this book. Without their support, this book would not have become a reality.

**Naser A. Anjum, Asim Masood and Nafees A. Khan**
Department of Botany,
Aligarh Muslim University,
Aligarh, India

**Shahid Umar**
Department of Botany,
School of Chemical and Life Sciences,
Jamia Hamdard University,
New Delhi, India

Chapter 1

# Introductory Chapter: Phosphorus in Soils and Plants

*Naser A. Anjum, Asim Masood, Shahid Umar and Nafees A. Khan*

## 1. Introduction

Plants (crops), soils, and humans are intricately interrelated. As heterotrophs, humans (and animals) largely depend on plants and/or on the plant-associated food resources. Seventeen (17) nutrient elements are varyingly required by plants in optimum level for the maintenance of their growth, metabolism, and productivity, as well as for their sustenance under adverse conditions. These elements are also called nutrients and include carbon (C), hydrogen (H), oxygen (O), nitrogen (N), phosphorus (P), potassium (K), sulfur (S), calcium (Ca), magnesium (Mg), boron (B), chlorine (Cl), copper (Cu), iron (Fe), manganese (Mn), molybdenum (Mo), nickel (Ni), and zinc (Zn). Interestingly, N, P, K, S, Ca, Mg, B, Cl, Cu, Fe, Mn, Mo, Ni, and Zn are derived from the soil (or parent rock materials) and can also be supplied by fertilizers. On the other hand, the list of non-minerals includes C, H, and O since these elements are obtained from water, air, or both. In addition to nourishing the plants, soils also provide anchorage, air ($O_2$), water, temperature modification, and the stabilization. As an essential plant macronutrient, P has been a key regulatory element for plant growth and metabolism. Unfortunately, P-bioavailability has been a worldwide constraint for plant growth and productivity. There must be a fine P-homeostasis in the soil-plant system in order to maintain the status of the soil-P and its efficient uptake and metabolism in plants, for optimum plant growth and productivity and eventually for attaining agricultural and environmental sustainability and the global food security.

## 2. Phosphorus in soils

Phosphorus (P; atomic number 15; nonmetal; placed in the 3rd row of the periodic of the elements; discovered by German chemist Hennig Brand in 1669) is a chemical element that occurs as white P and red P, its most common allotropic forms. It is never found as a free element on Earth and generally occurs as phosphates ($PO_4^{3-}$). P has 22 known isotopes which range from $^{26}P$ to $^{47}P$ [1]. Notably, $^{31}P$ is the only stable isotope of P and is therefore present at 100% abundance. Apatite (comprising minerals pentacalcium triorthophosphate fluoride) partly constitutes inorganic phosphate rock, the chief global commercial source of P, and a nonrenewable resource for the phosphate fertilizers.

Ranked 11th in the list of most abundant elements in the environment, P is neither easily accessible nor evenly distributed in most soils. In addition, the availability of the soil-P to plants/crops is limited. The concentration of P in the soil solution can be very high ($10^{-4}$ M), very low ($10^{-6}$ M), or as low as $10^{-8}$ M in some very poor tropical soils. Though P is not available in gaseous form, it is available in both inorganic and organic forms in the soil. In general, soil-P is available in both organic form (about 30–65%) and inorganic form (about 35–70%). Notably, plant-available (soil solution) P, sorbed P, and mineral P are the major pools of inorganic P-form. Contingent to the pH, $H_2PO_4^-$, $HPO_4^{2-}$, and $PO_4^{3-}$ are the major P-forms in the soil solution. P exhibits very low solubility and poor mobility in soil solution, as well as its capacity to form insoluble salts with different mineral elements. Soils with low amounts of organic matter and low water holding capacity are usually nutrient-deficient acidic soils widely reported to exhibit the least mobility of P as well as the least availability to plants (compared to N and K). P is also incorporated into organic compounds, where about 20–80% of soil-P is present in organic matter mainly as phytic acid (inositol hexakisphosphate), which can bind to various anions resulting in the formation of phytate. Most soil-organic P pools (mostly composed of plant and microorganism residues and livestock manures) cannot be absorbed directly by most plants [2]. Hence, P is among the less accessible elements for most plants [3]. $H_2PO_4^-$, denoted by inorganic phosphate (orthophosphate; Pi), dominates in the pH range of 3–7 and is the predominant form absorbed by plants. On the other hand, ester derivatives represent the organic P present in organic molecules. Phosphates are not reduced; however, the oxidized form of P is incorporated in the biomolecules [4]. There are several factors (physical, chemical, and biotic) known to regulate P-availability in soils and to the plant roots for its uptake. Major P-availability-modulating physical factors include soil texture and moisture; whereas chemical factors can be the soil solution pH, organic matter, redox potential, P-concentration in the soil solution, P-buffer capacity of the soil, and the concentration of Fe, Al, and Ca. On the other hand, major biotic factors influencing the plant/crop root activity and the P-release in the rhizosphere include the diversity of microorganisms [5–8].

## 3. Phosphorus in plants

In terms of its importance in most organisms, P has been considered an essential, irreplaceable element in all living cells. P stands second to N in terms of its essentiality as a macronutrient in plants/crops. P has been widely considered as a key regulatory element for plant growth and metabolism. Apart from providing an anchor for the plant in the soil, roots take up water and nutrients from the soil solution. In fact, the interception of nutrients available in the soil solution occurs in the rhizosphere. Eventually, the movement of nutrients toward the root involves mass flow and diffusion, where the former process contributes only 2–3% of the total amount of P (transported), usually required by many crops to produce acceptable yields. Interestingly, Pi, the only form of P that can be assimilated by plant, exhibits uneven distribution, relative immobile (as its diffusion coefficient is very low: $10^{-12}$ to $10^{-15}$ $m^2$ $s^{-1}$), and high fixation. Hence, the concentration of Pi is very low (<2.0 μM) in soil solution, and even in fertile soil (10 μM). In plants, P represents 0.1–0.5% of the dry weight. Therefore, Pi-availability has been a worldwide constraint for crop growth and productivity [3].

The knowledge of the plant Pi-acquisition and distribution mechanism may help in enhancing the plant/crop P-use efficiency. The list of the major strategies adopted by most plants/crops under P-deficiency includes: (i) increase in root surface area by formation of finer roots, aerenchyma, and root hairs, and eventual improved soil exploration; (ii) secretion of organic acids to release Pi through complexation reaction of organic acids with $Al^{3+}$, $Fe^{3+}$, and $Ca^{2+}$; (iii) the complexation reaction of organic acids with phosphatases to release Pi from the organic sources and thus to enhance the availability of Pi in soils; (iv) arbuscular mycorrhizal fungi (AMF) colonization; and (v) employing Pi transporters to facilitate Pi-uptake [3, 9, 10]. Arbuscular mycorrhizal fungi (AMF; phylum Glomeromycota) contribute to plant P-nutrition. Notably, AMF-colonization process is modulated by Pi-status and PHT1 transporter. Interestingly, this AMF-colonization is usually not well developed in soils with adequate, plant-available P. Both terrestrial plant species (90%) and in particular vascular plants, including main crops (>80%), exhibit AMF-colonization. Molecular genetic studies have unveiled various Pi-transporters, which facilitate the plant uptake and translocation of available Pi. Five phylogenetically distinct classes of families of plant Pi-transporters are known: PHT1, PHT2, PHT3, PHT4, and PHT5. PHT1s are, in general, plasma membrane-located and mainly function in Pi acquisition from soil. PHT2, PHT3, PHT4, and PHT5 family members are known to contribute in Pi distribution within the plant such as translocation against chloroplasts, mitochondria, Golgi, and vacuole. PHT1 transporters, the best studied plant Pi transporters, belong to multigene family and are involved in transport of Pi from apoplast/soil solution to the root cell through the plasma membrane [11]. Under P-deficiency/starvation, P-starvation response (PSR) pathway is activated, which involves PHOSPHATE TRANSPORTER TRAFFIC FACILITATOR1 (PHF1) and PHT1, and several transcription factors including PHOSPHATE STARVATION RESPONSE 1 (PHR1) and PHR1-LIKE 1 (PHL1), WRKY74, WRKY75, MYB2, MYB4, and ARF16 have been associated with the PSR pathway and implicated in P uptake [2, 12, 13]. Organic esters (such as phosphomonoesters and diesters) are the organic forms of P in plants. Phytase-mediated release of phosphates from phytates is possible during seed germination. Organic forms of P (namely, phosphomonoesters and diesters) are broken in plants by enzymes, namely, phosphatases and diesterases, in order to make P available to plants [4]. Plant responses to the low P (or P-starvation) also involve several hormones and signaling molecules (including cytokinins, CKs; abscisic acid, ABA; gibberellin, GA; and strigolactones, SLs) [14].

Interestingly, essential biomolecules such as DNA, RNA, ATP, NADPH, and membrane phospholipids involve P. In fact, nucleic acids, phosphorylated proteins, various phosphorylated metabolites, and phospholipids present the major pool of organic-P. Hence, plant tissues exhibit relatively high (about 5–20 mM) phosphate concentration. Phosphatases, nucleases, and phosphoesterases have been reported to contribute in the release of Pi from the organic forms during senescence in plants. Additionally, photosynthesis, respiration, and activation of proteins *via* phosphorylation, the life-sustaining processes in plants, strictly involve P. P is irreplaceable in many P-containing biomolecules. ATP and NADPH are the major high energetic molecules and are prerequisites to ensure photosynthesis functioning during the Calvin cycle [15]. Literature is full on the role of P in plant cell metabolism. The regulatory mechanisms during the developmental processes as well as response to stress conditions in plants involve phosphorylation of proteins. Moreover, intermediates in Calvin–Benson cycle are phosphorylated with Pi. Pi is released in the cytosol and is cycled back to plastids. The availability of Pi regulates photosynthesis, which is inhibited with limited Pi-supply. A plethora of literature supports the significant roles of P in imparting plant/crop tolerance to various abiotic stresses including drought [16–18], heavy metals [19–22], soil salinity [23–26], temperature [27–30], and water logging [31, 32].

## 4. Environmental concerns

The productivity of many agroecosystems is largely modulated by plant P-nutrition. Unfortunately, mainly as a result of phosphate extraction, fertilizers application, wastes generation, and P-losses from cropland, humans have perturbed the global P-cycle, which has tripled the global P-mobilization in land–water continuum and also increased P-accumulation in most soils [33]. Owing to the importance of P to life, and the considering the soil to water-sourced P as the major contributor for eutrophication in surface water bodies (and also in natural terrestrial habitats), the maintenance of the P-status of most world soils and its efficient use in agriculture have been focused in recent years. In particular, to attend the prevailing very low availability of Pi for plants/crops, intensive fertilization of crop plants with P-fertilizers (mainly as N-P-K fertilizers globally) is being adopted. Unfortunately, only 10–20% of applied Pi can be absorbed by plants, and rest can remain in the soils as legacy P. Thus, mainly intensive fertilization of crop plants with P-fertilizers has both impeded phosphate rock (non-renewable resource); and has also increased the fertility status of natural waters (eutrophication) due to legacy P (past P surpluses) in soils, sediments, and wastes. Elevated P levels in eutrophic water bodies cause increased growth of algae and large aquatic plants, leading to decreased levels of dissolved $O_2$, and/or algal toxin (produced by algae blooms) and eventual severe impacts on the health of aquatic biota, as well as that of humans and animals. The feeding of the crops with P but not the soils has been suggested in order to properly manage P in the food chain [34]. Thus, unscientific use of fertilizers has eaten away most soil nutrients, disturbed the soil-P balance, inhibited healthy plant growth, and provoked eutrophication and thus has become a major concern in sustainable agriculture production systems worldwide.

## 5. Conclusion and prospects

Among several nutrient elements that are varyingly required by plants, P is the major macronutrient involved in the maintenance of growth, metabolism, and productivity of plants, as well as for their sustenance under adverse conditions. Unfortunately, P is among the less accessible elements for plants mainly as a result of its very low solubility and poor mobility in soil solution, as well as its incorporation into organic compound. The availability of soil-P is modulated by several physical factors (soil texture and moisture), chemical factors (soil solution pH, organic matter, redox potential, and the concentration of Fe, Al, and Ca), and biotic factors (diversity of microorganisms in rhizosphere). Most plants/crops under P-deficiency adopt a range of strategies to facilitate plant/crop P-use efficiency. Pi-uptake and root-to-shoot transport are mediated by both low- and high-affinity Pi-transporters, whereas Pi-redistribution at tissue, cellular, and subcellular levels is facilitated by a dedicated sets of transporters. In addition to performing key functions in plants (such as energy transfer, photosynthesis, transformation of sugars and starches, and nutrient movement within the plant), an adequate P-supply help plants to sustain, develop, and produce under varied abiotic stress factors.

The knowledge of P-recycling and -mobilization may help in improving plant/crop P-use efficiency and thereby minimizing the P-fertilizer input to the environment. In particular, appropriate fertilizer encompassing both the change in the amount and suitability of P-fertilizer, as well as deciding the rate, time, and place of P-fertilizer, may help in achieving sustainable crop production. Exhaustive molecular genetic studies on P-solubilizing microorganisms, understanding Pi-transporters in tonoplast and Pi-uptake by the cells not symplastically connected to guard cells or the developing embryo, and identifying proteins mediating Pi-transport across tonoplast will be promising in the research area aimed at unveiling insights into P in soils and plants.

## Author details

Naser A. Anjum[1*], Asim Masood[1], Shahid Umar[2] and Nafees A. Khan[1]

1 Department of Botany, Aligarh Muslim University, Aligarh, India

2 Department of Botany, School of Chemical and Life Sciences, Jamia Hamdard, New Delhi, India

*Address all correspondence to: naanjum1@myamu.ac.in

## References

[1] Neufcourt L, Cao Y, Nazarewicz W, Olsen E, Viens F. Neutron drip line in the Ca region from Bayesian model averaging. Physical Review Letters. 2019;**122**(6):062502

[2] Nussaume L, Kanno S, Javot H, Marin E, Pochon N, Ayadi A, et al. Phosphate import in plants: Focus on the PHT1 transporters. Frontiers in Plant Science. 2011:2011, 2

[3] Wang D, Lv S, Jiang P, Li Y. Roles, regulation, and agricultural application of plant phosphate transporters. Frontiers in Plant Science. 2017;**8**:817

[4] Lal MA. Sulfur, phosphorus, and iron metabolism in plants. In: Physiology P, editor. Development and Metabolism. Springer: Singapore; 2018. pp. 481-515

[5] Jing Z, Chen R, Wei S, Feng Y, Zhang J, Lin X. Response and feedback of C mineralization to P availability driven by soil microorganisms. Soil Biology and Biochemistry. 2017;**105**:111-120

[6] Horta C, Roboredo M, Carneiro JP, Duarte AC, Torrent J, Sharpley A. Organic amendments as a source of phosphorus: Agronomic and environmental impact of different animal manures applied to an acid soil. Archives of Agronomy and Soil Science. 2018;**64**:257-271

[7] Young EO, Ross DS. Phosphorus mobilization in flooded riparian soils from the Lake Champlain Basin, VT, USA. Frontiers in Environmental Science. 2018;**6**:120

[8] Wang Y, Lambers H. Root-released organic anions in response to low phosphorus availability: Recent progress, challenges and future perspectives. Plant and Soil. 2020;**447**:135-156

[9] López-Arredondo DL, Leyva-González MA, González-Morales SI, López-Bucio J, Herrera-Estrella L. Phosphate nutrition: Improving low-phosphate tolerance in crops. Annual Review of Plant Biology. 2014;**65**:95-123

[10] Vengavasi K, Pandey R. Root exudation potential in contrasting soybean genotypes in response to low soil phosphorus availability is determined by photo-biochemical processes. Plant Physiology and Biochemistry. 2018;**124**:1-9

[11] Młodzińska E, Zboińska M. Phosphate uptake and allocation—A closer look at *Arabidopsis thaliana* L. and *Oryza sativa* L. Frontiers in Plant Science. 2016;**7**:1198

[12] Prathap V, Kumar A, Maheshwari C, Tyagi A. Phosphorus homeostasis: Acquisition, sensing, and long-distance signaling in plants. Molecular Biology Report. 2022;**49**:8071-8086

[13] Lapis-Gaza HR, Jost R, Finnegan PM. Arabidopsis PHOSPHATE TRANSPORTER1 genes PHT1;8 and PHT1;9 are involved in root-to-shoot translocation of orthophosphate. BMC Plant Biology. 2014;**2014**:14

[14] Khan F, Siddique AB, Shabala S, Zhou M, Zhao C. Phosphorus plays key roles in regulating plants' physiological responses to abiotic stresses. Plants. 2023;**12**(15):2861

[15] Hashida S, Kawai-Yamada M. Inter-organelle NAD metabolism underpinning light responsive NADP dynamics in plants. Frontiers in Plant Science. 2019;**10**:960

[16] Tariq A, Pan K, Olatunji OA, Graciano C, Li Z, Sun F, et al.

Phosphorous application improves drought tolerance of *Phoebe zhennan*. Frontiers in Plant Science. 2017;**8**:1561

[17] Razaq M, Zhang P, Shen H, Salahuddin S. Influence of nitrogen and phosphorous on the growth and root morphology of *Acer mono*. PLoS One. 2017;**12**:e0171321

[18] Begum N, Ahanger MA, Zhang L. AMF inoculation and phosphorus supplementation alleviates drought induced growth and photosynthetic decline in *Nicotiana tabacum* by up-regulating antioxidant metabolism and osmolyte accumulation. Environmental and Experimental Botany. 2020;**176**:104088

[19] Li L, Xing W, Ma C, Zhang Y, Wang G, Yang L. Phosphorus amendment of a lead-spiked soil with low phosphorus availability: Roles of phosphorus on soil and plant lead. Communications in Soil Science and Plant Analysis. 2012;**43**:1053-1064

[20] Gomes MP, Carvalho M, CarvalhoGS,MarquesTCLLSM,GarciaQS, Guilherme LRG, et al. Phosphorus improves arsenic phytoremediation by *Anadenanthera peregrina* by alleviating induced oxidative stress. International Journal of Phytoremediation. 2013;**15**:633-646

[21] Zeng G, Wan J, Huang D, Hu L, Huang C, Cheng M, et al. Precipitation, adsorption and rhizosphere effect: The mechanisms for phosphate-induced Pb immobilization in soils - a review. Journal of Hazardous Materials. 2017;**339**:354-367

[22] Cao X. Immobilization of heavy metals in contaminated soils amended by phosphate-, carbonate-, and silicate-based materials: From lab to field. In: Yongming L, Tu C, editors. Twenty Years of Research and Development on Soil Pollution and Remediation in China. Singapore: Springer; 2018. pp. 535-543

[23] Miranda R, de Ruppenthal V, Lopes LS, Fontenele C, Marques VB, et al. Phosphorus fertilization improves soybean growth under salt stress. Frontiers in Plant Science. 2012;**10**:1116

[24] Assaha DVM, Ueda A, Saneoka H, Al-Yahyai R, Yaish MW. The role of $Na^+$ and $K^+$ transporters in salt stress adaptation in glycophytes. Frontiers in Physiology. 2017;**8**:509

[25] Tang H, Niu L, Wei J, Chen X, Chen Y. Phosphorus limitation improved salt tolerance in maize through tissue mass density increase, osmolytes accumulation, and $Na^+$ uptake inhibition. Frontiers in Plant Science. 2019;**10**:856

[26] Bouras H, Bouaziz A, Bouazzama B, Hirich A, Choukr-Allah R. How phosphorus fertilization alleviates the effect of salinity on sugar beet (*Beta vulgaris* L.) productivity and quality. Agronomy. 2021;**11**:1491

[27] Fahad S, Hussain S, Saud S, Hassan S, Ihsan Z, Shah AN, et al. Exogenously applied plant growth regulators enhance the morpho-physiological growth and yield of rice under high temperature. Frontiers in Plant Science. 2016;**7**:1250

[28] Fahad S, Bajwa AA, Nazir U, Anjum SA, Farooq A, Zohaib A, et al. Crop production under drought and heat stress: Plant responses and management options. Frontiers in Plant Science. 2017;**8**:1147

[29] Singh SK, Reddy VR, Fleisher DH, Timlin DJ. Phosphorus nutrition affects temperature response of soybean growth and canopy photosynthesis. Frontiers in Plant Science. 2018;**9**:1116

[30] Xu H, Hassan MA, Sun D, Wu Z, Jiang G, Liu B, et al. Effects of low temperature stress on source-sink organs in wheat and phosphorus mitigation strategies. Frontiers in Plant Science. 2022;**13**:807844

[31] Ayub M, Nadeem MA, Sharar MS, Mahmood N. Response of maize (*Zea mays* L.) fodder to different levels of nitrogen and phosphorus. Asian Journal of Plant Science. 2002;**1**:352-354

[32] JdlC J, Cardoso JA, Arango-Londoño D, Fischer G, Rao I. Influence of soil fertility on waterlogging tolerance of two *Brachiaria* grasses. Agronomia Colombiana. 2015;**33**:20-28

[33] Yuan Z, Jiang S, Sheng H, Liu X, Hua H, Liu X, et al. Human perturbation of the global phosphorus cycle: Changes and consequences. Environmental Science and Technology. 2018;**52**(5):2438-2450

[34] Withers PJ, Sylvester-Bradley R, Jones DL, Healey JR, Talboys PJ. Feed the crop not the soil: Rethinking phosphorus management in the food chain. Environmental Science and Technology. 2014;**48**(12):6523-6530

Chapter 2

# Phosphorus Dynamics in Soil-Water-Sediment Environment

*Eisa Ebrahimi and Mohammad Reza Ojani*

## Abstract

Phosphorus, a crucial element for plant growth, is relatively scarce in the Earth's crust. Its availability in surface soils ranges from 100 to 50 mg/kg. Plants can only absorb phosphorus in the form of orthophosphates, with $H_2PO_4$- being most absorbable at low pH levels. The average phosphorus concentration in soil solution is about 0.05 mg/L, but plant-satisfying levels range from 0.003 to 0.3 mg/L, underscoring the need for judicious phosphorus fertilization. Storage and stabilization reactions in soil, mainly facilitated by iron and aluminum oxides, play a key role. Compounds in most soils use hydroxyl exchange mechanisms for $H_2PO_4$- adsorption. Under alkaline conditions, minerals like calcium carbonate can absorb $H_2PO_4$-/ $HPO_4^{2-}$, leading to precipitation. To ensure plant health, phosphorus fertilizers, especially calcium orthophosphates like triple superphosphate, are commonly used. These fertilizers offer essential phosphorus for plant growth and development, supporting vital processes like respiration and photosynthesis. Excess phosphorus in aquatic ecosystems, known as eutrophication, poses environmental risks, often originating from concentrated agricultural and livestock operations. Proper management of phosphorus inputs is crucial for balancing plant growth support with environmental preservation. Understanding phosphorus dynamics in soil, water, and sediments is vital for sustainable agriculture and conservation efforts. Adsorption isotherms provide insights into phosphorus absorption mechanisms in sediments, impacting water quality in surface and subsurface systems.

**Keywords:** phosphorus, orthophosphates, phosphate rocks, eutrophication, trophic states, isotherms

## 1. Introduction

Phosphorus is a non-metallic element in the $5^{th}$ group of the periodic table with an atomic weight of 30.97, a density of 1.82 g/cm$^3$, and an atomic number of 15. It was discovered in 1669 by a German chemist named Hennig Brand. The melting point and boiling point of this element are 44.1 and 280°C, respectively, and its atomic radius is 1.28 angstroms. Phosphorus is never found in nature in its free state; instead, it's found combined with other elements in rocks, phosphate minerals, and all living cells. It appears as a waxy white solid with an unpleasant smell. Phosphorus is highly reactive and emits a small amount of light when it reacts with oxygen. It burns rapidly in air, transforming into phosphorus pentoxide. This non-metal does not dissolve in water, but it is soluble in carbon disulfide [1]. Phosphorus is an essential component of biological molecules such as ATP, ADP, and DNA, and it's also found in bones and teeth. It is present in trace amounts in all the foods we eat. There are various isotopes of phosphorus; only one, 31P, is stable. The other isotopes of phosphorus are radioactive and have short half-lives ranging from a few nanoseconds to a few seconds, except for two isotopes, 32P and 33P, which have longer half-lives of 14 and 25 days, respectively. These longer half-lives make these isotopes useful for analyzing and labeling DNA.

## 2. Phosphorus in soil

Phosphorus is an essential element for plant growth and survival. However, the amount of phosphorus available in the Earth's crust is relatively low compared to other essential plant nutrients, ranging from 1100 to 1200 mg/kg. The average concentration of phosphorus in surface soils is estimated to be around 1100–50 mg/kg. Plants can only absorb phosphorus in the form of orthophosphates, with the highest concentration being in the form of $H_2PO_4^-$ at low pH levels. This form of orthophosphate is more readily absorbed by plants compared to $HPO_4^{2-}$, which increases in concentration with increasing pH levels. The average concentration of phosphorus in the soil solution is approximately 0.05 mg/L, although this value can vary depending on the soil type. Typically, the concentration of phosphorus in the soil solution that can satisfy plant needs ranges from 0.003 to 0.3 mg/L, as reported by [2, 3]. These findings suggest that while phosphorus is essential for plant growth and survival, it is often limited in availability in the soil. Therefore, it is crucial to carefully manage phosphorus fertilization practices to ensure that plants have access to the necessary amounts of this nutrient for optimal growth and development.

Phosphorus in soil exists in both organic and mineral forms. Typically, 20–80% of total soil phosphorus is in the organic form, while the remainder is in mineral form [4, 5]. However, these forms are not strictly distinct from one another and can transform into each other (**Figure 1**). The specific types of organic phosphorus are not well understood, but mineral phosphorus is typically associated with iron, aluminum, and calcium. These minerals each have varying solubilities in the soil. The mobility and speciation of mineral phosphorus in soil and aquatic ecosystems are regulated by reactions that occur on the surfaces of colloids [6–8]. Additionally, Lindsay noted that other reactions, such as simultaneous precipitation by calcium, iron, and aluminum, can also influence the phosphorus cycle in soil [9, 10].

The mechanisms responsible for converting soluble phosphorus into less soluble states are related to phosphorus storage and stabilization reactions. Nearly all soils contain iron and aluminum oxides and hydroxides, which can exist as separate mineral particles or as coatings on other particles, such as clay. Amorphous aluminum hydroxides may also be present between the expandable layers of aluminosilicates. Many $H_2PO_4^-$ adsorbents are present in the soil solution, and most of these adsorbents, such as iron or aluminum, use the hydroxyl ($OH^-$) exchange mechanism to absorb $H_2PO_4^-$. In alkaline conditions, minerals like calcium carbonate can absorb $H_2PO_4^-/HPO_4^{2-}$ and cause precipitation by exchanging water, bicarbonate, and hydroxyl ions on their surfaces. Notably, each of these reactions occurs in response to the soil's pH level. At pH levels lower than 5, hydroxyl metal phosphates (Al $(OH)_2H_2PO_4$) can form due to the presence of aluminum, iron, or active manganese.

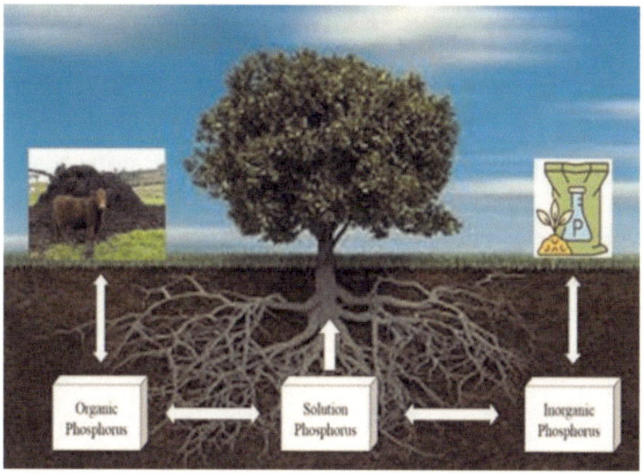


**Figure 1.**
*Phosphoric and mineral cycle in soil.*

In alkaline conditions, dicalcium phosphate deposits without water molecules can form due to the presence of active calcium [6, 11, 12].

As previously mentioned, the natural concentration of phosphorus in the soil is very low, and this small amount of phosphorus is not easily available to the plant. The plant has to compete with the reactions of surface absorption and phosphorus deposition in the soil to obtain the phosphorus it needs. Therefore, to achieve acceptable performance, phosphorus fertilizers should be added to the soil to aid the growth and nutrition of the plant. The most common phosphorus fertilizers are calcium orthophosphate fertilizers, including triple superphosphate and simple superphosphate. Triple superphosphate, with the formula $Ca(H_2PO_4)_2$, has 44–53% phosphorus oxide [2, 13, 14].

## 2.1 Sources of phosphorous input

### *2.1.1 Phosphorous fertilizers*

The growth and health of plants depend on several factors, including nutrient supply, soil quality, and exposure to sunlight. Among these, phosphorus is a crucial element that plays a significant role in promoting plant growth and health. Despite its importance in supporting respiration, photosynthesis, cell division and enlargement, and energy storage and transfer, the amount of phosphorus in soil is typically much lower than that of other essential elements like potassium, nitrogen, and calcium. As a result, various types of phosphorus-based fertilizers are produced today to supplement the soil's phosphorus content and help improve plant growth.

In the nineteenth century, researchers began conducting experiments to examine the effects of phosphorus fertilizers on plant yield. Britain (1843) and Germany (1878) were among the pioneering countries in this field, with subsequent experiments conducted at the Illinois Agricultural Research Station in 1888. Further experiments were carried out in Ontario, Canada, in 1916 [15]. After World War II, the use of animal manure to maintain soil fertility and provide food for the growing population began to expand rapidly [16].

### *2.1.2 Sources of phosphorus fertilizers*

#### *2.1.2.1 Phosphate rocks*

These are a valuable source of phosphorus for plants, with the most reactive rocks containing Francolite, an Apatite mineral containing iron and carbonate. When crushed, phosphate rock can provide sufficient plant-available phosphorus in low pH soils. It is commonly used to cultivate plants such as Kathira, oil palm, and coffee in very acidic soils, particularly in warm weather conditions, moist soils, and long growing seasons.

#### *2.1.2.2 Phosphoric acid*

Phosphoric acid ($H_3PO_4$) is another important source of phosphorus for agriculture, containing 17–24% phosphorus (55–39% $P_2O_5$). It is produced through the reaction of phosphate rock with $H_2SO_4$ and can be either green acid or acid obtained from a wet process. Phosphoric acid is used to acidify phosphate rock and create calcium and ammonium phosphates, which are important fertilizers. It can also be injected into the soil or irrigation water, particularly in alkaline and calcareous soils.

#### *2.1.2.3 Calcium phosphates*

In the past, calcium phosphate fertilizers such as simple superphosphate, triple superphosphate, and enriched superphosphate were among the primary sources of phosphorus. However, unlike phosphoric acid and ammonium phosphates, superphosphates do not have a significant impact on soil pH.

#### *2.1.2.4 Simple superphosphate*

Also known as single superphosphate, it contains 7–9–5% phosphorus ($P_2O_5$ 16–22%). It is a valuable source of both sulfur and phosphorus.

#### *2.1.2.5 Triple superphosphate*

This contains 17–23% phosphorus ($P_2O_5$ 44–52%). Triple Super Phosphate is an excellent source of phosphorus, and its high concentration of phosphorus is advantageous because transportation, storage, and handling constitute a significant part of the total cost of fertilizer consumption.

#### *2.1.2.6 Ammonium phosphate*

Ammonium phosphate is produced by reacting phosphoric acid with $NH_3$. Monoammonium phosphate (MAP) contains 11–13% nitrogen and 21–24% phosphorus ($P_2O_5$ 48–55%), while diammonium phosphate (DAP) contains 18–21% nitrogen and 20–23% phosphorus ($P_2O_5$ 53–46%). Both MAP and DAP are granular and water-soluble, making them suitable as starter fertilizers. Their high solubility allows for efficient uptake by plants, resulting in improved growth and yield.

#### *2.1.2.7 Ammonium polyphosphate*

Ammonium polyphosphate is produced by reacting pyrophosphoric acid with ammonia. Liquid APP is a cost-effective source of phosphorus that can be used alone or in combination with other liquid fertilizers. Typically, UAN and APP fertilizers are combined and applied in bands below the soil surface.

#### *2.1.2.8 Potassium phosphate*

Potassium phosphate products, such as $KH_2PO_4$ and $K_2HPO_4$, are highly soluble in water and are commonly used in the horticulture industry. Their high levels of both potassium and phosphorus make them a popular choice for crops such as potatoes, tomatoes, and many leafy vegetables that are sensitive to high concentrations of chloride found in KCl. Additionally, their low salt index makes them well-suited for application near seeds with minimal damage to emerging sprouts.

## 3. Phosphorous in water

In the 1960s, it was discovered that aquatic ecosystems were experiencing high levels of phosphorus. Further scientific investigations revealed that this phosphorus was originating from livestock farms and concentrated agriculture located kilometers away from these water ecosystems. These areas, which are the source of phosphorus input to aquatic ecosystems, are known as concentrated polluting points. During that decade, due to advancements in technology and the creation of laws limiting the risk of pollution from these points, identifying and controlling these sources became easier. However, today, the transfer of phosphorus from unknown agricultural sources is a serious and worrying threat to the environment, as it is practically impossible to control these sources. The phosphorus eventually reaches water reservoirs, causing many problems. This results in exorbitant fees for using these waters [17, 18].

Phosphorus is lost from watersheds due to soil erosion and washing and is transferred to aquatic ecosystems through surface and subsurface runoff. When the runoff reaches surface water, the washed (dissolved) elements along with the eroded materials are emptied into these waters, creating certain problems. Sediments fill the reservoirs of dams, corrode water treatment facilities, and reduce the quality of drinking water. The elements present in these waters cause the phenomenon of enrichment, which cannot be ignored because it can cause many problems. Water enrichment causes toxic substances to be released from aquatic organisms into the water, leading to diseases in humans, with neurological problems being the most common in recent decades. The transfer of phosphorus to surface waters, deposition, and release of phosphorus from sediments into the water causes the phenomenon of surface water enrichment [19–21].

Phosphorus losses are typically observed in the following three ways in the field [22]:

a. Dissolved phosphorus and/or phosphorus bound to solid soil particles-suspended in water-created by erosion and accompanied by runoff flows, which are transferred to surface waters such as rivers and lakes.

b. Dissolved phosphorus and/or phosphorus bound to solid soil particles reach the surface of underground water sources due to infiltration runoff in the soil profile (Leaching).

c. Seepage or lateral infiltration of phosphorus transferred to underground water to lakes, rivers, and other surface streams.

On the other hand, the transfer of phosphorus from topsoil to surface and subsurface waters can be summarized in three mechanisms:

## 3.1 Sources of phosphorus input

Water pollution refers to the contamination of lakes, rivers, oceans, and underground waters by chemical and microbial waste products. When harmful substances are directly or indirectly discharged into water sources, and if these substances limit the widespread use of water, the water is considered polluted. The effects of water pollution are felt by plants and living organisms that inhabit these waters. In many cases, the effects of pollution destroy natural biological groups as well as individual and collective species [23].

Non-point pollution can cause physical damage to waterways by increasing sedimentation. This occurs when erosion deposits fine soil particles in the water, making it murky. The extent of sediment accumulation is influenced by factors such as the speed of the flow, shear stress, and soil grain stability. The type and quantity of sediment that originates from agricultural areas differ from that of forested areas [24]. The consequences of rising sediment levels are serious, including:

- Silt particles in the water can wear out the water pump and other facilities related to water transfer.

- Using murky water in agriculture can also cause problems. For instance, utilizing muddy water in irrigation systems leads to wear and tear of pumps, distribution systems, and other irrigation devices. Consequently, this can result in complications with the irrigation system.

- The introduction of suspended substances into streams, lakes, and reservoirs threatens aquatic life. Studies have shown that the greater the amount of suspended matter in the water, the more detrimental the effects on aquatic life.

- Sediments can cloud the water, preventing sufficient light from reaching aquatic depths, thereby disrupting the photosynthesis process.

- The contamination of waters with mud reduces the esthetic appeal of recreational centers.

### *3.1.1 Point and non-point sources*

Water pollutants are divided into two categories based on the emission source: point pollutant (point or concentrated pollution) and non-point pollutant source (non-point pollution).

A point source of pollution refers to a specific and identifiable location that releases pollutants into the receiving environment. Examples of such sources include wastewater discharged by industries, power plants, and urban sewage treatment facilities. The Clean Water Act (CWA) in the United States provides a regulatory definition for point sources. The CWA's definition of a point source was updated in 1987 to encompass municipal sewage systems, industrial wastewater, and construction wastewater.

A non-point pollutant source refers to a source of pollution that does not have a specific and identifiable entry point into the receiving environment. Typically, non-point sources include water and runoff from agricultural lands, mines, construction sites, roads, and urban areas. Additionally, air pollution settling on water sources also constitutes non-point pollution. Unlike point source pollution, which originates from a single source, non-point pollution is the result of the accumulation of pollution amounts collected from a large basin. Non-point sources of pollution pose the biggest threat to surface and subsurface drinking water sources worldwide. Erosion is also a significant source of non-point pollution due to the large amounts of chemical fertilizers, insecticides, pesticides, and other substances used in agricultural lands, which have an average erosion rate higher than other land uses. Consequently, agricultural lands are one of the primary sources of this type of pollution.

### *3.1.2 Soil*

Phosphorus is a crucial element for modern agriculture, widely used in chemical fertilizers to enhance crop yields. In recent years, phosphorus consumption has increased due to the depletion of soil phosphorus caused by extensive crop harvesting. The significance of phosphorus in the crop production system is evident from the doubling of its use in fertilizers since 1960, while its annual global production has remained less than 2 million tons over the past decade. As depicted in **Figure 2**, phosphorus primarily leaves the land through runoff in solution and attached to particles, while a small fraction infiltrates the groundwater through preferential flow. The potential for phosphorus to reach groundwater is relatively low compared to nitrogen, owing to its low solubility. However, soil particles have a high capacity to adsorb insoluble forms of phosphorus, and hence, soil functions like a filter. Phosphorus loss from agricultural land is classified into three levels:

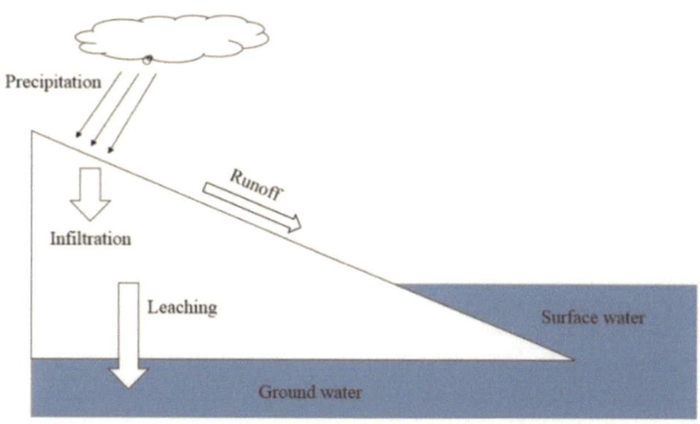


**Figure 2.**
*Potential routes of phosphorus exit from agricultural lands.*

a. Rapid loss of soluble phosphorus in a short period of time after the application of animal and chemical fertilizers.

b. Slow loss of soluble phosphorus after fertilization.

c. Loss due to erosion.

Animal and chemical fertilizers contain a significant amount of soluble phosphorus, which can enter runoff due to rainfall after application and increase the concentration of this element in the runoff up to 100 times its normal level [25]. The amount of phosphorus that can be stored or adsorbed on the surface of clay depends on the type and quantity of clay, iron oxide, and organic matter. The critical concentration of phosphorus varies in different surface waters, as shown in **Table 1** [26].

Phosphorus is primarily released from agricultural lands in association with sediments. Proper management of plowing can reduce erosion and phosphorus loss. Therefore, in agricultural fields with a steep slope, it is advisable to use protective plows or perform contour cultivation [27]. Studies have demonstrated that total phosphorus losses are higher in fields with Chisel plowing compared to minimum plowing, whereas soluble phosphorus losses are higher in fields with minimum plowing [28]. The retention or release of phosphorus in water depends on the physical, chemical, and biological conditions of the compounds in the water system. The intensity of phosphorus discharge varies from one river to another, and even within different parts of a river, changing with seasonal variations [29]. Phosphorus storage in water is influenced by several factors, such as the biomass of macrophytes, the absorption of this element by aquatic plants, the expansion of the river margin, and nutrient concentration [30].

### *3.1.2.1 Phosphorus leaching*

Some researchers suggest that phosphorus can be absorbed into the subsoil through the vertical movement of water in the soil profile, as the subsoil layers have a high absorption capacity for phosphorus [31]. Thus, the transfer of phosphorus from these soils to underground water and drains is expected to be very low or negligible [24]. However, studies have shown that despite the high absorption capacity of the underlying soils, significant concentrations of phosphorus can still be found in the drains of these fields [11]. These studies indicate the presence of preferential flow,

| Phosphorous concentration (μg/L) | Explanation | Reference |
|---|---|---|
| 10 | Average dissolved phosphorus in runoff from agricultural fields | Vollenweider (1968) |
| 100 | Critical concentration of dissolved phosphorus in lakes | USEPA (1986) |
| 50 | Critical concentration of total phosphorus in current | USEPA (1986) |
| 1000 | Critical concentration of total phosphorus in lakes | USEPA (1986) |

*United States Environmental Protection Agency*

**Table 1.**
*Critical concentration of phosphorus in surface waters.*

which is a flow that occurs through permanent gaps in the field, and may develop more during dry seasons [32]. The creation of preferential flow does not require certain conditions such as the intensity and duration of rainfall or the permeability and saturation conditions needed to create surface runoff [33]. It should also be noted that preferential flow reduces the contact time between the soil and the infiltrating water, further contributing to the transfer of phosphorus to the drains. Even though subsurface soils have a high absorption capacity, preferential flow can still result in the transfer of phosphorus from these soils to the drains.

#### *3.1.2.2 Erosion and runoff*

Soil erosion and its impact on surface water is a complex process that depends on various factors such as soil characteristics, hydrological, and climatic conditions. Farm sediment erosion, which occurs in furrows and inter-furrows, is influenced by raindrop impact and surface currents. Raindrops typically cause soil particles to detach from each other, while surface currents cause both particle detachment and transport down the slope. The effectiveness of these processes is greatly affected by surface soil type, rainfall duration, and the amount of plant cover and residues. Soil surface coverage and erosion have an inverse relationship, so implementing soil conservation practices can reduce erosion [34, 35]. Runoff usually occurs in fields through two types of flow: surface and subsurface flow. Surface flow can sometimes penetrate into the soil along the slope and move laterally below the surface before reappearing on the soil surface. Losses of phosphorus in agricultural land occur through surface runoff, which can be in the form of dissolved or sediment-bound phosphorus. Phosphorus bound to sediments includes phosphorus that is bound with soil mineral particles and organic matter. During surface runoff in cultivated fields, 80% of phosphorus losses occur in the form of sediment-bound phosphorus [36]. In contrast, runoff from grasslands, meadows, uncultivated soils, and forests transports fewer sediments, so most of the phosphorus that is transported downstream is in the form of dissolved phosphorus in water [24, 37]. During runoff movement along sloping surfaces, phosphorus from the soil and plant residues can dissolve and enter the runoff (**Figure 3**). In summary, [39] have concluded in their research that the total phosphorus loss can be determined based on the sources of phosphorus, the chemical and biological changes and transformations of phosphorus, and the mobility of phosphorus in the soil.

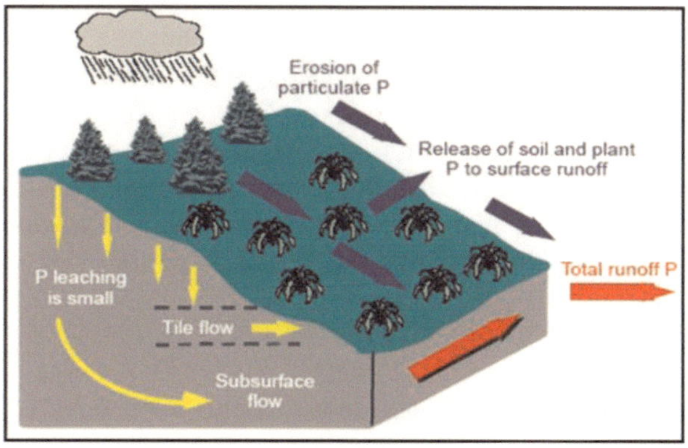


**Figure 3.**
*Phosphorus movement paths in the field (adapted from [38]).*

The interaction between the surface soil (1–2 inches) and water (rain or irrigation water) creates the possibility of releasing phosphorus in a soluble form from the soil and plant residues. In most watersheds, the amount of phosphorus transferred through surface runoff is greater than subsurface runoff, as phosphorus is typically low in deep soil layers. When runoff penetrates into the soil, water-soluble phosphorus is absorbed into the subsurface soil, reducing the concentration of phosphorus in the subsurface runoff compared to surface runoff. However, sandy soils, peaty soils, and soils with preferential flow can exclude this case [40].

The subsurface transfer of phosphorus has been studied very little, and it was previously assumed that the amount of transfer through subsurface runoff is very small. However, some researchers believe that a significant portion of phosphorus (more than 0.1 mg/L) is transported in the form of orthophosphate in subsurface runoff [41]. Although other forms of phosphorus may also be found in subsurface runoff [33], studies have reported a correlation between the amount of dissolved phosphorus in the runoff and the amount of phosphorus in the soil. If the amount of phosphorus in the soil exceeds 60 mg/kg, the concentration of phosphorus in the runoff will significantly increase [32]. The type of vegetation and management practices can also have a significant impact on this phenomenon.

When the concentration of phosphorus exceeds 60 mg/kg, the concentration of total phosphorus in the runoff increases sharply. The reason for this is that when phosphorus exceeds this limit, it is absorbed into sites with less absorption and storage energy and is easily released into water [42]. However, using arsenic phosphorus to predict the concentration of dissolved phosphorus in runoff has many problems, including the depth of sampling to determine arsenic phosphorus, which is usually taken from the depth of the plow. The depth of the interaction between the runoff and the soil is estimated to be between 1 and 2 inches [24]. To overcome such problems, two Dutch scientists, [43], used the soil phosphorus saturation percentage to assess the risk of phosphorus leaching into underground water. This percentage is calculated by dividing the amount of phosphorus available to the plant by the maximum amount of phosphorus fixation in the soil. During their research, a critical limit of 25% for phosphorus saturation was established (for soils in the Netherlands). If the phosphorus saturation exceeds this amount, phosphorus will be released into the water. Every soil has the potential to absorb and store phosphorus, and this potential depends on the physical-chemical characteristics of the soil. The importance of iron and aluminum oxides and other soil minerals in controlling the solubility and absorption of phosphate in the soil has been identified [2].

The rate of weathering and release of phosphorus from phosphate rocks is very low and is estimated to be about 0.01–5 km/ha.year [44]. The rate of weathering and release of phosphorus from phosphate rocks depends on effective factors such as the type of phosphorus rock, their size, temperature, and water quality. A positive correlation exists between the amount of chemical weathering and runoff generated in a watershed. When the amount of runoff is high, the amount of weathering is almost 10 times more than when the amount of runoff is low [45].

### *3.1.3 Sediment*

Sediment moves downstream along slopes through waterways. Larger particles are deposited first due to the reduction in runoff transmission power, while finer particles, such as organic materials and clays, along with silt-sized particles, enter water ecosystems along with water-soluble phosphorus. Eventually, these sediments reach rivers and water reservoirs, such as natural lakes, ponds, and reservoirs behind dams, and settle. Some of the eroded materials are so small that they settle much more slowly and can be resuspended with even slight turbulence in the water. Dynamic water conditions or changes in the redox potential can resuspend solid particles and release their adsorbed elements, such as phosphorus and nitrogen, into the water [46–48]. Therefore, the relationship between suspended particles and dissolved phosphorus in lake water is of great chemical importance.

Four main mechanisms have been identified for the release of phosphorus from sediments [49–51]. These mechanisms can cause changes in the mineral structure of sediments, which can, in turn, lead to the release of phosphorus into the water column above the sediment [46].

The four mechanisms are briefly discussed hereunder.

Absorption: Phosphorus desorption can occur in the following three different situations:

a. Phosphorus bound to calcium and magnesium minerals may be released into the water due to a reduction in pH [52].

b. An increase in temperature can also affect pH. In this case, phosphorus may be separated from iron, aluminum, or manganese [53].

c. The increase in the concentration of sulfate and chlorate ions in water may compete with phosphate ions for absorption on iron oxides [54].

Dissolution: The dissolution of minerals containing phosphorus may cause the release of this element under the following conditions:

a. Microbial activities can produce organic or mineral acids that can dissolve phosphorus bound to iron, aluminum, or manganese [55]. These activities may also release chelates such as gluconate and 2-ketogluconate. Other factors to consider include the revival of $H_2S$ and iron. Additionally, a decrease in pH can result from the absorption of ammonium ions by fungi, leading to dissolution [56].

b. Nitrate increases the redox potential [56]. Other cases include the recovery of sulfate to pyrite and the recovery of methane.

c. Stumm and Morgan [54] pointed out that an increase in salinity and a decrease in pH decrease phosphorus absorption on iron oxides.

Ligand exchange: As a result of increasing pH, the concentration of $OH^-$ increases and ligand exchange takes place with $PO_4^{-3}$ located on iron hydroxides [57].

Enzymatic hydrolyses: Enzymatic hydrolyses in aquatic ecosystems are critical processes driven by microbial activities.

#### *3.1.3.1 Microbial activities*

Enzymes present in the structure of microbes cause mineralization of organic phosphorus and accelerate its release [14, 58]. An increase in temperature due to the activity of microbes intensifies microbial activities [53].

#### *3.1.3.2 Decomposition*

During decomposition, a recovery state occurs. In this case, trivalent iron is regenerated to ferric iron, and phosphorus bound to iron is released into the sediments. Manganese also undergoes regeneration under these conditions [59].

## 3.2 Eutrophication

The term "eutrophic" originates from the Greek word "Eutrophos," meaning "good formation" and "proper union." Similarly, the term "eutrophication" is derived from the Latin words "affectus," meaning "good," and "nutrimens," meaning "food." In Greek, "trophi" means food, and this term is usually accompanied by prefixes such as "oligo," "meso," "eu," and "hyper," which mean "rare, little," "moderate," "abundant," and "severe, high," respectively. Therefore, the words "oligotrophic," "mesotrophic," "eutrophic," and "hypertrophic" are used to express the different states of nutrient levels in lakes or freshwater environments.

There are several definitions of eutrophication, and some of these definitions have fundamental differences from others. One of the differences is related to whether this phenomenon is simply a mechanism for increasing nutrients or if it should also include issues related to enrichment [60–62].

a. In South Africa, this phenomenon is generally defined as "eutrophication is the natural aging mechanism of lakes" [60].

b. The natural mechanism of increasing biological production due to the increase (enrichment) of nutrients [63].

c. Excessive growth of algae due to the enrichment of water with nutrients, especially nitrogenous and phosphorous compounds, and the reduction of water quality due to the disruption of the existing balance between microorganisms [62].

### *3.2.1 Eutrophication factors*

In natural lakes, there is a distinction between natural and artificial (created by humans) eutrophication. Natural eutrophication depends only on the geology of the area and the natural characteristics of the watershed. If eutrophication is related to human activity (adding a large amount of nutrients to the aquatic ecosystem), which exacerbates this natural phenomenon, these nutrients originate from point and non-point sources of pollution, such as phosphorus in sediments [60]. The change in land use alters the amount of nutrients in the runoff. According to studies conducted, phosphorus removed from agricultural lands is at least five times greater, and from urban areas, 10 times greater than from forest lands. Enrichment of water by nutrients exists in nature, but human activities intensify it, and this process occurs almost everywhere in the world. Three important human-related sources of nutrient input are:

A. Runoff and erosion.

B. Leaching of substances in the chemical fertilizers from agricultural fields.

C. Wastewater from urban areas and sewage from factories.

The most significant cause of eutrophication is the introduction of a large amount of nutrients into the water, which causes an imbalance in the nutrient cycle and greatly increases the amount of phytoplankton biomass, ultimately leading to the growth of algae. The direct result of the increase in biomass is the high consumption of oxygen underwater [62].

### 3.2.2 Trophic states

Various indicators are used to measure the degree of this phenomenon, and one of the best indicators is the use of Nurenberg's classification [64]. This classification is based on the concentration of total phosphorus and nitrogen and chlorophyll a (**Table 2**) [65, 66].

The concept of trophic state is complex and lacks a precise definition. In the past, the trophic state was mainly related to the amount of nutrients in a lake. However, it now also takes into account biological factors that can cause changes in the lake's morphology. The term "eutrophic state" is not only applicable to lakes with high nutrient levels, but also to shallow wetlands and places where aquatic plants grow, even if nutrient levels are not high. Lakes are classified into four groups based on their trophic state: "oligotrophic," "mesotrophic," "eutrophic," and "hypertrophic."

#### 3.2.2.1 Oligotrophy

If the water is in this state, it contains few nutrients and there is minimal biological activity. The level of photosynthesis is very low and the water is very clear, allowing sunlight to penetrate deeply. Usually, the dissolved oxygen content in the water is high.

#### 3.2.2.2 Eutrophy

In eutrophic conditions, the situation becomes more complex. In this case, the water contains an abundance of nutrients, which fuels high levels of biological activity. As a result, light is absorbed on the surface of the water and does not reach the depths, resulting in cloudy water. Oxygen levels in the water are very low, and deposited organic materials decompose through anaerobic processes, further contributing to eutrophication. The lack of oxygen in these waters also makes them

| Trophy status | Total phosphorus concentration | Total nitrogen concentration | Chlorophyll a concentration |
|---|---|---|---|
| Oligotrophic | >0.01 | >0.35 | >0.0035 |
| Mesotrophic | 0.03–0.01 | 0.65–0.35 | 0.009–0.0035 |
| Eutrophic | 0.1–0.03 | 1.2–0.65 | 0.025–0.009 |
| Hypertrophic | 0.1< | 1.2< | >0.025 |

*All units are based on mg/l.*

**Table 2.**
*Proposed classification system for the status of the Nuremberg trophy.*

uninhabitable for fish, which often die off. Nutrient enrichment is the main factor responsible for the degradation of surface water quality, causing intense growth of algae, which can reduce activities related to fisheries, recreation, and industry.

Algae grow rapidly in eutrophic waters and form a layer on the surface that can lead to a reduction in dissolved oxygen levels, increased suspended solids, and a decrease in the diversity of aquatic species. Strong growth of cyanobacteria, in particular, can occur in these waters, causing health problems for humans and livestock if these waters are used. The growth of algae can also release toxic and volatile substances, causing nerve damage and raising concerns about the phenomenon of nutrient enrichment. While nitrogen and carbon are also essential for the growth of aquatic organisms, controlling their levels is difficult due to the atmospheric and water cycles of these elements. However, phosphorus has been identified as the most limiting element for the growth of aquatic organisms, as evidenced by numerous studies (**Figure 4**).

### *3.2.3 Consequences of eutrophication*

The primary consequence of eutrophication is a decrease in the amount of dissolved oxygen in water. During photosynthesis, plants release oxygen into the water using sunlight. Conversely, in dark conditions, organisms, plants, and aerobic microorganisms consume dissolved oxygen during respiration. The balance between photosynthesis and respiration depends on the growth and population of the biomass. As a result of the high accumulation of biomass and the presence of an oxidation mechanism, underwater sediments form from the living mass, leading to a depletion of dissolved oxygen. Additionally, some decomposing bacteria consume oxygen from sulfate, resulting in the release of sulfur and the trapping of oxygen in the upper layers of the water. This reduction in oxygen levels endangers aquatic life. In the absence of oxygen, certain types of food can become toxic compounds, such as nitrate turning into ammonium, sulfate turning into hydrogen sulfide, and carbon dioxide turning into methane. These compounds are harmful to aquatic organisms.

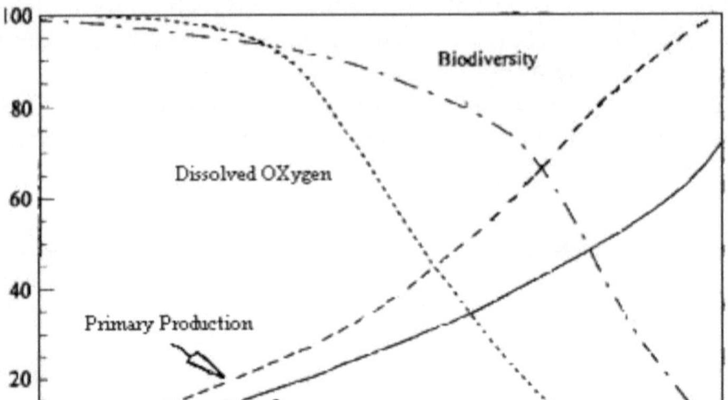


**Figure 4.**
*The role of phosphorus in eutrophication of water.*

A. Increasing the cost and problems associated with purifying water to make it drinkable.

B. Decreasing the quality of edible fish and replacing them with fast-growing macrophytes and algae.

C. Accumulation of sediment on the seabed, which causes a decrease in depth, an increase in vegetation, and disturbance to navigation.

D. Decomposition of algae products and surface scum that cause unpleasant odors (hydrogen sulfide, methane, etc.).

E. Increasing the population of pathogenic insects and mosquitoes.

### 3.3 Evidences on the importance of phosphorus in aquatic ecosystems

Experiments have shown that in lake systems, an increase in phosphorus relative to nitrogen leads to a significant increase in the number of cells of certain algae. This suggests that phosphorus is more limiting than nitrogen in this type of lake [67]. Additionally, evidence has shown that in cases of phosphorus deficiency in aquatic ecosystems, there is a direct correlation between the growth rate of algae (the rate of cell division) and the amount of phosphorus available within each cell [68, 69].

In 1972, Powers et al., conducted an experiment in which they placed 320 liters of water from Lake Organ and Minnesota in a closed environment and enriched it with various nutrients. By adding phosphorus, they observed a positive response from the system and concluded that phosphorus is the primary limiting factor. An experiment conducted by Schindler between 1974 and 1977 in the Ontario experimental lake clearly demonstrated the limitation of phosphorus. In this experiment, the lake water was saturated with phosphorus for several years, and the lake used atmospheric carbon and nitrogen to grow algae. The result of this work was an increase in primary nutrients, creating a eutrophic state in the lake water. An excess of phosphorus in the water acted as the trigger for the excessive bloom of cyanobacteria. When phosphorus, carbon, and nitrogen were added in deficiency, the effects were minimal [70].

In 1976, Vollenweider Weider created a model that could predict the eutrophic state in lakes and water reservoirs, with the only input element being phosphorus. This model was used worldwide and predicted the creation of the eutrophic state with very high accuracy. This model is in the form of Eq. 1:

$$Cl_a = \frac{(L_p/Q_s)}{\left[1 + \sqrt{\frac{z}{Q_s}}\right]} \tag{1}$$

These are: $Cl_a$ is the algal biomass in units (mg/m$^3$), $L_p$ is the input amount of phosphorus (g/m$^2$.d$^1$) and $Q_s$ is the output amount of the lake per unit of lake area (m/Acre).

A group of researchers has discovered a sigmoidal relationship between the logarithm of the total phosphorus concentration in the summer and the logarithm of chlorophyll levels. As the total phosphorus concentration increases, the amount of chlorophyll reaches a constant state, beyond which the amount of chlorophyll remains almost constant with further increases in phosphorus. This finding contrasts with the previously assumed linear relationship. In the case where the phosphorus concentration reaches its maximum value, adding more phosphorus to the water does not increase the chlorophyll concentration, while adding nitrogen intensifies the enrichment phenomenon [71]. It should be noted that as we move from freshwaters to coastal waters and oceans, the limiting element changes from phosphorus to nitrogen [72]. However, some researchers have challenged this idea [73]. Due to the high persistence of phosphorus in lakes, it is still considered the most limiting factor [70].

### 3.4 Different parts of phosphorus in water

There are three main terms used to describe phosphorus in water: soluble reactive phosphorus (DRP), total phosphorus (TP), and phosphorus bound to suspended solids. DRP refers to the portion of water phosphorus that can pass through a filter with a pore size of less than or equal to 0.45 μm and is analyzed using a colorimetric method. This method only measures the readily available portion of phosphorus, known as soluble orthophosphate, which directly contributes to water enrichment for aquatic plants. However, it should be noted that the filter with a pore size of less than 0.45 micrometers sometimes fails to effectively separate phosphorus.

To obtain the total phosphorus content of water, unfiltered water is digested with a strong acid, and the phosphorus content is measured using a colorimetric method. During the digestion stage, polyphosphates and phosphates attached to organic materials, which cannot be measured using the colorimetric method, are converted to orthophosphate and measured. Subtracting DRP from TP yields the phosphorus bound to suspended solids in the water. It is important to note that suspended solids play a crucial role in the phosphorus cycle in water. Therefore, water analysis measures these three parts of phosphorus: DRP, phosphorus bound to suspended solids, and TP (which is the sum of the previous two parts).

### 3.5 Phosphorus cycle in water and the role of suspended particles and sediments in this cycle

In aquatic ecosystems such as lakes and rivers, phosphorus is typically present in the pentavalent form. This includes various types of compounds such as orthophosphates, pyrophosphates, long-chain polyphosphates, organophosphate esters, and phosphodiesters, as well as organic phosphates. When phosphorus enters surface water, it can become attached to solid particles in the form of phosphates and organophosphates. Through chemical or enzymatic processes, this bound phosphorus may then be released into the water as orthophosphates, which are the only form of phosphorus that can be absorbed by plants, algae, and bacteria through hydrolysis. Phosphorus is highly dynamic in water environments and is biologically active. Once phosphorus reaches surface water, it may become attached to solid particles and settle at the bottom of lakes. Microbial communities can then gradually consume the organic parts of these particles, releasing the phosphorus in the form of orthophosphate back into the water, as shown in **Figure 5**. It is important to note that phosphorus bound to solid particles and dissolved organic phosphorus are not inert in

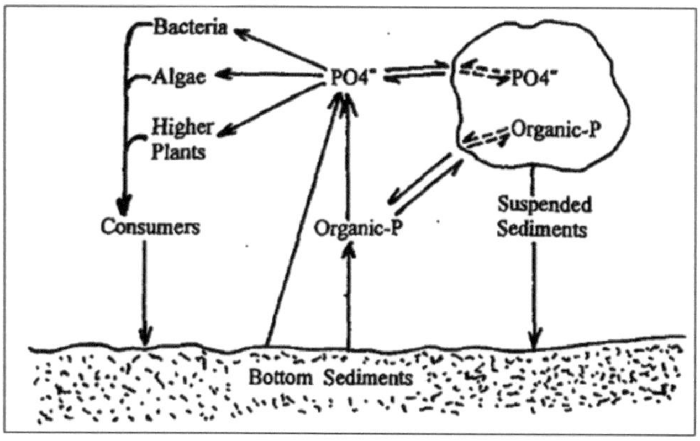


**Figure 5.**
*Schematic diagram of the phosphorus cycle between water and bottom sediments [70].*

water conditions, as they can convert into soluble orthophosphate under certain conditions.

Phosphorus that enters lakes, reservoirs, and river estuaries is absorbed by various biological structures, sediments, and living organisms and deposited in the depths (as shown in **Figure 5**). This process helps prevent the widespread distribution of phosphorus in the water, but it also renders aquatic ecosystems highly sensitive to excessive phosphorus. In oligotrophic conditions, the deposited phosphorus may remain in sediments for years. However, in eutrophic conditions, the deep water is often in an anaerobic state, which can also occur in shallow waters on warm, windless nights. In this eutrophic state, most of the phosphorus deposited in sediments decomposes and spreads into the water.

Some researchers suggest that phosphorus attached to solid particles entering surface waters does not have much effect on the growth of aquatic algae [74]. However, several reports have indicated that dissolved orthophosphates in surface waters are linked to their dynamic relationship with particulate phosphorus and sediments at the lake bottom [75]. The dynamic balance between phosphorus bound to suspended particles and dissolved phosphorus is known as a phosphate buffering mechanism in water [76]. From a kinetic perspective, dissolved phosphorus reacts with suspended particles in water during two stages. The first stage, or fast stage, occurs in less than a few seconds, while the second stage, or slow stage, takes several days. During the fast phase, phosphorus attaches to the surfaces of suspended particles, but during the slow phase, phosphorus penetrates into the structure of the particles. When runoff with suspended particles is discharged into surface waters, exchanges occur between phosphorus attached to these particles and phosphorus dissolved in the water, creating a new balance. If the concentration of soluble phosphorus in the lake water is low, phosphorus is released from the suspended sediments into the water, and vice versa.

As suspended particles settle on the lake bottom, complex conditions arise. Bacterial activities cause gradual mineralization of organic phosphorus, releasing it into the water within the pores of lake bottom sediments. In the next step, dissolved phosphorus may spread into the lake water or be absorbed by the surface of the particles (sediments) before it can spread into the water. Phosphorus bonding with aluminum hydroxides and ferric iron on the surface of sediments can be very strong. If the water in the sediments' pores undergoes regeneration (due to biological activities), ferric iron is converted to ferrous iron, and the bond between iron and phosphorus becomes very weak, facilitating phosphorus to enter the water column [50, 77, 78]. Therefore, the amount of phosphorus exchange between the water column and bottom sediments may change seasonally.

## 3.6 Range of phosphorus soluble in water

To understand the extent of eutrophication caused by phosphorus in water, it is crucial to measure the total phosphorus concentration instead of just orthophosphate. This is because orthophosphate concentrations can rapidly change within minutes under enriched conditions. Laboratory studies indicate that the concentration of phosphate required for balanced algae growth ranges from 0.003 to 0.8 μg/L. However, in Lake Michigan, it was observed that a concentration of 15 μg/L significantly increased carbon fixation and chlorophyll concentration. While a few micrograms per liter of phosphorus can enhance algae growth in most water systems, there is no agreement among researchers on the exact concentration that causes eutrophication. Some researchers suggest that a concentration between 0.01 and 0.03 mg/L of phosphorus can lead to excessive growth of harmful algae, while others propose a threshold of 0.02 mg/L for accelerated eutrophication. It is essential to note that the appropriate threshold for phosphorus concentration depends on the specific area being studied.

## 3.7 Methodologies for addressing insights into phosphorus pollution in aquatic environments

Various methodologies are employed to study and address phosphorus pollution in aquatic environments.

These methods allow researchers and environmental professionals to assess the extent of pollution, identify sources, and develop effective mitigation strategies.

Here are the brief details of some common methodologies used.

### *3.7.1 Water sampling and analysis*

Water samples are collected from different points within a water body and analyzed for phosphorus content. Techniques like colorimetry, spectrophotometry, and inductively coupled plasma (ICP) analysis are used to quantify total phosphorus and different forms (e.g., dissolved, particulate). This helps in understanding the distribution and concentration of phosphorus in the water column. Here's how water sampling and analysis are applied to the study of phosphorus pollution in aquatic systems:

#### *3.7.1.1 Sample collection*

Water samples are collected from different locations, depths, and time points within the aquatic environment. Sampling may target areas influenced by pollution sources, such as agricultural runoff or wastewater discharges.

#### *3.7.1.2 Phosphorus forms*

Different forms of phosphorus are analyzed, including dissolved reactive phosphorus (DRP), particulate phosphorus, and total phosphorus. These forms provide insights into the various pathways and sources of phosphorus pollution.

#### *3.7.1.3 Sampling depth profile*

Collecting samples at different depths allows researchers to understand vertical variations in phosphorus concentrations. This is particularly important in stratified water bodies.

#### *3.7.1.4 Sampling frequency*

Frequent sampling over time provides data on temporal variations in phosphorus levels, capturing seasonal changes and short-term fluctuations.

#### *3.7.1.5 In-situ measurement*

In addition to collecting samples for laboratory analysis, in situ sensors and probes can measure real-time water quality parameters, including phosphorus levels.

#### *3.7.1.6 Laboratory analysis*

Laboratory techniques such as colorimetry, spectrophotometry, and ICP analysis are used to quantify phosphorus concentrations in water samples. These methods provide accurate and quantitative data.

#### *3.7.1.7 Quality control*

Proper sample handling, preservation, and storage are essential to maintain the integrity of collected samples. Quality control measures ensure that results are reliable.

#### *3.7.1.8 Nutrient ratios*

Besides phosphorus concentrations, nutrient ratios such as the nitrogen-to-phosphorus ratio (N:P ratio) are often assessed to understand nutrient limitations and potential impacts on ecosystem dynamics.

#### *3.7.1.9 Source identification*

Isotopic analysis of phosphorus can help identify pollution sources. Different sources have distinctive isotopic signatures that can be traced in water bodies.

#### *3.7.1.10 Spatial mapping*

Results from water sampling can be spatially mapped using Geographic Information Systems (GIS) to visualize the distribution of phosphorus pollution within aquatic systems.

*3.7.1.11 Long-term monitoring*

Continuous or periodic water sampling and analysis provide data for long-term monitoring programs, allowing researchers to track changes over years or decades.

*3.7.1.12 Baseline assessment*

Water sampling establishes baseline data on phosphorus levels, aiding in evaluating the effectiveness of pollution control measures and management strategies.

*3.7.1.13 Regulatory compliance*

Water sampling is often conducted to assess compliance with water quality standards and regulations related to phosphorus pollution.

*3.7.2 Sediment analysis*

Sediments can act as reservoirs of phosphorus, contributing to its release into the water. Sediment core sampling and analysis provide information about historical phosphorus deposition and storage. Sequential extraction methods are used to determine different forms of phosphorus bound to sediments.

*3.7.3 Nutrient budgets*

Nutrient budgets involve accounting for inputs and outputs of phosphorus within a water body. This includes quantifying point sources (e.g., wastewater treatment plants) and non-point sources (e.g., agriculture runoff). By understanding the sources and pathways, effective management strategies can be devised.

*3.7.4 Modeling*

Mathematical models, such as hydrodynamic and water quality models, simulate phosphorus transport and distribution in aquatic systems. These models consider factors like flow rates, sedimentation, and nutrient interactions to predict phosphorus behavior under different scenarios. Modeling and mathematical approaches help synthesize complex interactions within aquatic systems and provide insights into how phosphorus pollution affects water quality, ecosystem health, and the efficacy of management strategies. They are valuable tools for designing effective interventions to mitigate phosphorus pollution and its environmental consequences.

*3.7.4.1 Hydrodynamic models*

These models simulate water movement, flow patterns, and circulation within aquatic systems. They help understand how phosphorus is transported through water bodies and how it interacts with sediment and other components.

*3.7.4.2 Water quality models*

Water quality models integrate hydrodynamics and biochemical processes to simulate nutrient dynamics, including phosphorus concentrations. These models consider factors like nutrient uptake, sediment interactions, and biological processes that influence phosphorus levels.

#### *3.7.4.3 Eutrophication models*

Eutrophication models specifically focus on nutrient enrichment, including phosphorus, and its effects on aquatic ecosystems. They help predict the development of algal blooms, oxygen depletion, and other consequences of nutrient pollution.

#### *3.7.4.4 Mass balance models*

Mass balance models quantify phosphorus inputs, outputs, and internal cycling within a water body. They consider sources such as point and non-point pollution, as well as sinks like sedimentation and nutrient uptake by organisms.

#### *3.7.4.5 Reaction kinetics models*

These models describe the rates at which chemical reactions involving phosphorus occur in aquatic systems. They help predict how phosphorus transformations and interactions with other elements change over time.

#### *3.7.4.6 Sediment transport models*

These models focus on sediment dynamics and how they transport phosphorus. They help understand erosion, sedimentation, and the release of phosphorus from sediment into the water column.

#### *3.7.4.7 Nutrient loading models*

Nutrient loading models estimate the amount of phosphorus entering aquatic systems from various sources, such as agricultural runoff, urban stormwater, and wastewater discharge.

#### *3.7.4.8 Scenario analysis*

Modeling allows researchers to explore different scenarios and assess the potential outcomes of management strategies. For example, they can predict the effects of reducing phosphorus inputs from specific sources.

#### *3.7.4.9 Validation and calibration*

Models are often calibrated and validated using real-world data to ensure accuracy. This involves adjusting model parameters to match observed conditions and using historical data to test the model's predictive capabilities.

#### *3.7.4.10 Policy and decision support*

Mathematical models provide a scientific basis for making informed decisions about phosphorus management. They help evaluate the potential impact of different policies and management actions.

*3.7.4.11 Data integration*

Models often require input data such as nutrient concentrations, flow rates, and bathymetry. Remote sensing, water sampling, and other data collection methods provide essential information for model inputs and validation.

## 3.8 Phosphorus estimation methodologies

*3.8.1 Estimation of different forms of phosphorus*

Phosphorus exists in various forms, including dissolved reactive phosphorus (DRP), particulate phosphorus (PP), and total phosphorus (TP). Laboratory methods are crucial for measuring different forms of phosphorus in water. There are a number of methods widely used for measuring different forms of phosphorus in water samples. However, it is important to note that the choice of method depends on factors such as the specific form of phosphorus being measured, the sensitivity required, and the laboratory's equipment and capabilities. Proper sample collection, handling, and preservation are essential to ensure accurate results. Additionally, quality control measures, including calibration with standards and replicates, help ensure the reliability of measurements.

Here are common laboratory methods used to measure these different forms:

*3.8.1.1 Dissolved reactive phosphorus (DRP)*

*3.8.1.1.1 Molybdate blue method*

This colorimetric method involves reacting phosphate ions with molybdate reagents to form a blue complex. The intensity of the blue color is proportional to the concentration of DRP and can be measured spectrophotometrically.

*Major procedure involved include:*

a. Sample collection and preparation

- *Collect a representative water sample in a clean and acid-washed container to prevent contamination.*
- *If necessary, filter the sample to remove any particles that could interfere with the colorimetric reaction.*

b. Reagent preparation

- *Prepare the molybdate reagent by mixing ammonium molybdate and a reducing agent (ascorbic acid) in an acidic solution.*
- *The reagents are usually available as a kit or can be prepared according to standard protocols.*

c. Color development

- *In a cuvette or test tube, add a measured volume of the sample and an appropriate volume of the molybdate reagent.*

- *Allow the mixture to react for a specific period to allow the formation of the molybdenum blue complex.*

d. Measurement

- *Using a spectrophotometer, measure the absorbance of the blue-colored solution at a specific wavelength, typically around 880 nm.*
- *The absorbance value is proportional to the concentration of DRP in the sample.*

e. Calibration and calculation

- *Create a calibration curve using known concentrations of phosphate standards treated with the same reagents and procedure.*
- *Use the calibration curve to determine the concentration of DRP in the sample based on its absorbance value.*

f. Quality control

- *Include blank samples (water without phosphate) and replicate measurements to ensure accuracy.*
- *Regularly calibrate the spectrophotometer using standard solutions.*

g. Reporting results

- *Express the results in units of phosphorus concentration (e.g., mg/L or μg/L).*

*3.8.1.1.2 Ascorbic acid method*

Phosphate reacts with ascorbic acid in the presence of ammonium molybdate to form a blue color. The color intensity is measured using a spectrophotometer.

*Major procedure involved include:*

a. Sample collection and preparation

- *Collect a representative water sample and ensure it is properly preserved and stored to prevent changes in phosphorus concentrations.*
- *If needed, filter the sample to remove any particulate matter.*

b. Reagent preparation

- *Prepare the molybdate reagent by dissolving ammonium molybdate in an acidic solution.*
- *Prepare a solution of ascorbic acid in deionized water.*

c. Color development

- *In a cuvette or test tube, add a measured volume of the sample, the molybdate reagent, and the ascorbic acid solution.*

- *Mix the contents thoroughly and allow the reaction to occur for a specified period.*

d. Measurement

- *Use a spectrophotometer to measure the absorbance of the blue-colored solution at a specific wavelength, usually around 880 nm.*

e. Calibration and calculation

- *Create a calibration curve using known concentrations of phosphate standards treated with the same reagents and procedure.*
- *Use the calibration curve to determine the concentration of DRP in the sample based on its absorbance value.*

f. Quality control

- *Include blank samples (water without phosphate) and perform replicate measurements to ensure accuracy.*
- *Regularly calibrate the spectrophotometer using standard solutions.*

g. Reporting results

- *Express the results in units of phosphorus concentration, such as mg/L or μg/L.*

*3.8.1.2 Particulate phosphorus (PP)*

- Filtration and Gravimetric Method: Water samples are filtered to separate suspended particles. The filter is dried and weighed before and after filtration. The difference in weight represents the mass of particulate phosphorus.

*3.8.1.3 Total phosphorus (TP)*

a. Persulfate Digestion Method: Water samples are digested with persulfate to oxidize all forms of phosphorus to orthophosphate. The resulting orthophosphate is then measured using colorimetric methods like those mentioned for DRP.

*Major procedure involved include:*

b. Sample collection and preservation

- *Collect a representative water sample in a clean and acid-washed container. For accurate results, ensure proper preservation and storage.*

c. Reagent preparation

- *Prepare a reagent solution by dissolving ammonium persulfate and potassium persulfate in deionized water.*

- *Optionally, prepare a blank solution using deionized water to account for any background contamination.*

d. Digestion

- *In a digestion tube or vessel, add a measured volume of the water sample and an appropriate volume of the persulfate reagent.*
- *Heat the mixture in a digestion block or other appropriate heating apparatus to a high temperature (usually around 120–160°C) for a specified period (often about 1–2 hours).*
- *The high temperature breaks down the organic and inorganic phosphorus compounds, converting them to orthophosphate.*

e. Cooling and dilution

- *Allow the digested sample to cool to room temperature after the digestion period.*
- *Dilute the digested sample with deionized water to bring it within the working range of the colorimetric method.*

f. Orthophosphate measurement

- *Use a colorimetric method, such as the Molybdate Blue Method or the Ascorbic Acid Method, to measure the concentration of orthophosphate in the digested sample.*
- *Prepare standard solutions of known phosphate concentration to create a calibration curve for accurate quantification.*

g. Quality control

- *Perform replicate measurements and include a blank solution to ensure accuracy.*
- *Regularly calibrate the colorimetric instrument using standard solutions.*

h. Calculation and reporting

- *Calculate the total phosphorus concentration in the sample based on the measured orthophosphate concentration and any dilution factors used.*

### 3.9 Monitoring employing remote sensing

Satellite and aerial imagery can be used to detect changes in water quality associated with phosphorus pollution. Remote sensing allows for the monitoring of algal blooms, turbidity, and other indicators of nutrient enrichment. Remote sensing plays a crucial role in assessing and monitoring phosphorus pollution in aquatic environments. It provides valuable information about water quality, algal blooms, sediment dynamics, and other indicators of nutrient enrichment.

Here are some key applications of remote sensing in the context of phosphorus pollution in aquatic systems:

### *3.9.1 Algal bloom detection and monitoring*

Remote sensing can identify and track algal blooms, which often result from excessive phosphorus levels. Sensors on satellites and aircraft can detect the unique spectral signatures of chlorophyll-a, a pigment in algae. Monitoring changes in chlorophyll-a concentrations helps in early detection and management of harmful algal blooms.

### *3.9.2 Water transparency and turbidity*

Phosphorus pollution can lead to reduced water transparency due to suspended particles. Remote sensing measures water turbidity, which is indicative of sediment and nutrient loads. Monitoring turbidity helps understand how phosphorus affects water clarity.

### *3.9.3 Total phosphorus concentrations*

Remote sensing algorithms can estimate total phosphorus concentrations in water bodies by analyzing spectral data. Correlations between spectral reflectance and nutrient concentrations allow researchers to infer phosphorus levels.

### *3.9.4 Erosion and sediment transport*

Sediment transport is closely linked to phosphorus pollution. Remote sensing helps monitor changes in shoreline morphology and sediment plumes, providing insights into erosion patterns and sediment dynamics.

### *3.9.5 Land use and land cover changes*

Remote sensing helps monitor changes in land use and land cover, including urban expansion and agricultural activities. These changes often contribute to phosphorus pollution through runoff and erosion.

### *3.9.6 Reservoir management*

Remote sensing assists in managing reservoirs by tracking sediment deposition and nutrient accumulation. This information helps make informed decisions about water quality and ecosystem health.

### *3.9.7 Spatial analysis and mapping*

Remote sensing data, when combined with Geographic Information Systems (GIS), enables the creation of spatial distribution maps of phosphorus concentrations and pollution sources. These maps guide management efforts.

#### *3.9.8 Temporal trend analysis*

Remote sensing allows for long-term monitoring of phosphorus-related parameters. Analyzing trends over time helps understand the effects of phosphorus pollution and assess the effectiveness of mitigation strategies.

#### *3.9.9 Early warning systems*

Remote sensing can be integrated into early warning systems for algal blooms. Timely detection of blooms allows water managers to take proactive measures to protect public health and ecosystem integrity.

#### *3.9.10 Data integration*

Remote sensing data can be combined with other datasets, such as water quality measurements and meteorological data, to develop comprehensive models for predicting phosphorus pollution dynamics.

### 3.10 Monitoring employing biological monitoring

Aquatic organisms respond to changes in phosphorus levels. Monitoring the health and composition of aquatic communities can provide insights into the effects of phosphorus pollution. Biotic indices, like the Trophic State Index, are used to assess water quality based on the presence of indicator species.

### 3.11 Mitigation approaches

#### *3.11.1 Best management practices (BMPs)*

BMPs involve implementing strategies to reduce phosphorus inputs. This includes practices such as buffer zones, cover crops, and optimized fertilizer application in agriculture, as well as improved wastewater treatment methods. BMPs play a significant role in mitigating phosphorus pollution in aquatic environments. These practices are a set of guidelines, strategies, and techniques designed to minimize the release of phosphorus into water bodies. They address various pollution sources and promote sustainable land and water management.

Here are some key BMPs used to tackle phosphorus pollution in aquatic systems.

##### *3.11.1.1 Agricultural BMPs*

- Nutrient management plans: Implementing precise fertilizer application based on soil nutrient levels to reduce excess phosphorus runoff from agricultural fields.

- Cover crops: Planting cover crops during fallow periods to reduce soil erosion and nutrient runoff.

- Buffer zones: Creating vegetated buffers along water bodies to trap sediment and nutrients before they enter aquatic environments.

- Conservation tillage: Reducing soil disturbance during planting to minimize erosion and nutrient loss.
- Rotational grazing: Managing livestock grazing to prevent overgrazing, soil compaction, and nutrient runoff.

*3.11.1.2 Urban BMPs*

- Storm water management: Implementing green infrastructure, such as permeable pavements, rain gardens, and detention basins, to capture and treat stormwater runoff before it enters water bodies.
- Erosion control: Using erosion control measures on construction sites to prevent sediment and nutrient runoff into nearby water bodies.
- Reduced lawn fertilization: Encouraging responsible lawn care practices, including proper fertilization and avoiding excessive phosphorus use on lawns.

*3.11.1.3 Wastewater treatment BMPs*

- Advanced treatment systems: Upgrading wastewater treatment plants with advanced nutrient removal technologies to reduce phosphorus discharge into receiving waters.
- Phosphorus removal facilities: Adding specific treatment units to remove phosphorus from wastewater before discharge.
- Septic system maintenance: Properly maintaining septic systems to prevent nutrient leaching into groundwater and nearby water bodies.

*3.11.1.4 Riparian zone restoration*

- Vegetation planting: Restoring native vegetation along shorelines and riverbanks to stabilize soil, prevent erosion, and filter runoff.
- Natural shoreline design: Creating gentle, natural shorelines instead of hardened structures to promote sediment and nutrient retention.

*3.11.1.5 Educational and outreach programs*

- Public awareness: Educating residents, farmers, and businesses about the impacts of phosphorus pollution and the importance of BMPs.
- Training workshops: Providing training and workshops on implementing effective BMPs to different stakeholders.

#### *3.11.1.6 Land use planning and zoning*

- Zoning regulations: Incorporating phosphorus pollution considerations into land use planning and zoning regulations to prevent new pollution sources.

- Smart growth: Encouraging compact and sustainable development to reduce urban sprawl and its associated runoff.

#### *3.11.1.7 Research and monitoring*

- Monitoring programs: Establishing long-term water quality monitoring programs to track phosphorus levels and assess the effectiveness of BMPs.

- Adaptive management: Using collected data to adjust and refine BMP implementation based on real-world results.

### *3.11.2 Riparian restoration*

Restoring riparian vegetation along water bodies helps in reducing phosphorus runoff and erosion. Plant roots stabilize soil, preventing sediment and phosphorus from entering waterways.

### *3.11.3 Green infrastructure*

Incorporating green infrastructure like constructed wetlands and vegetated swales can capture and filter phosphorus from storm water runoff before it reaches water bodies.

### *3.11.4 Public awareness and education*

Increasing public understanding of the sources and consequences of phosphorus pollution can lead to better practices, such as proper waste disposal and reduced fertilizer use.

## 4. Phosphorus in sediments

The role of sediment as a chemical pollutant depends on various factors, such as the size of the particles that make up the sediment, and the amount of organic matter and nutrients present in it. Nutrients can be released through surface runoff, which is influenced by soil type, vegetation, rainfall, and land use practices. The use of phosphate and other fertilizers in agricultural lands, as well as landfills, can contribute to the transfer of persistent pollutants to the soil and, ultimately, to runoff through leaching [79].

Since erosion tends to selectively transport the smallest soil particles, which are often rich in nutrients and organic substances, the high specific surface area of these particles can lead to the absorption of nutrients, toxins, and pests, resulting in their accumulation in sediments. Suspended sediments are the primary vehicle for transporting nutrients, toxic substances, and chemical elements in aquatic environments. Therefore, sediments from eroded soils can have different environmental impacts depending on the materials and elements they contain.

Concerns about the increase in nitrogen, phosphorus, and pesticide compounds in surface and underground waters have been growing in Europe since the 1970s. Agriculture is identified as the main source of phosphorus compounds and sediments in a recent comparison between domestic, industrial, and agricultural sources of pollution in the Mediterranean basin. The European community has responded with the directive EEC 16/6/1991, which aims to protect waters from pollution caused by nitrates from agriculture. In France, this issue has led to the formation of an advisory committee under the supervision of the Ministry of Agriculture and Environment to reduce nitrogen and phosphorus pollution in agriculture.

An analysis by the United States Environmental Protection Agency in 1994 identified agriculture as the primary land use that degrades wetlands and other wetland areas. As nutrients are removed from the soil during the erosion process and transferred to rivers, water reservoirs, and other water sources, eutrophication and oxygen depletion can occur, leading to a decline in water quality [80–82].

The physical-chemical properties of sediments, including amorphous iron and aluminum content, the amount of clay, and other characteristics, play a critical role in the absorption of phosphorus. Once absorbed, phosphorus can be released back into the water due to various biological and chemical activities, leading to water enrichment in certain cases. As stated by Sposito [83], the absorption and removal of phosphorus from surfaces are more complex than for other elements in soil.

## 4.1 Adsorption of phosphorus on sediments

Adsorption of phosphorus onto sediments is subject to complex conditions, as the physical, chemical, and biological characteristics of sediments play decisive roles in the absorption and retention of phosphorus. Under different environmental conditions, each of these features plays a different role. Iron and aluminum oxides and hydroxides are among the important chemical properties known to play a role in phosphorus absorption and storage [17]. The high ability of metal oxides and hydroxides to absorb phosphorus is probably due to their high specific levels [84]. Research has shown that the bond between phosphorus and amorphous aluminum is stronger than that with iron, as the bond with aluminum involves both electrostatic and hydrogen interactions, whereas with iron, it is primarily electrostatic [85]. Conducted research on sediment drains and concluded that over 88% of the phosphorus absorption capacity is related to the iron extracted with ammonium oxalate [86].

The amount of silt and clay in sediments is another parameter known to be important in phosphorus absorption. Sanyal and De Datta [87] have highlighted the significance of the relationship between phosphorus absorption and the amount of clay, which may be due to the high specific levels of clay. Some researchers have shown that reducing the particle size exponentially increases the amount of phosphorus absorption [88, 89]. It should also be noted that smaller particles are more easily disturbed by factors such as water waves and become suspended, providing more opportunities to react with dissolved phosphorus in water compared to larger particles [90]. Phosphorus is absorbed in the form of an inner-sphere complex on colloid surfaces, meaning that phosphorus absorption takes place in specific positions on colloid surfaces, and no water molecule is present between the surface and phosphate ion. The phosphate ion is directly absorbed by the functional groups on the surface of colloids. Phosphorus absorption occurs in two stages: surface absorption and sedimentation in a sequential form [85].

Organic carbon is one of the most important parameters in absorption, and studies have shown that the absorption capacity of phosphorus is dependent on the amount of organic matter in the sediments [91, 92]. There is a significant relationship between the absorption capacity and the amount of organic matter [20]. Ngoyan and Sukias [86] reported a very high correlation between iron and aluminum extracted with oxalate and organic carbon. This study showed that iron and aluminum may bond with organic carbon and play a much more effective role in phosphorus absorption. Under calcareous conditions, simultaneous absorption or precipitation of phosphorus with calcium carbonate has been reported. Solid phase reactions of $CaCO_3$ control phosphorus reactions. The pH level of sediments also plays a key role in phosphorus absorption. With a change in pH, iron and aluminum precipitation occurs. When the pH is higher than 8, phosphorus precipitates by forming a bond with calcium [93].

### 4.2 Phosphorus adsorption isotherms

Previous studies on soil have shown that the process of phosphorus absorption is non-linear due to varying energy levels in different binding sites on particles. The high-energy sites are usually occupied and filled first. The exchange of phosphorus between sediments that are resuspended in water and the surrounding water follows a two-stage dynamic, with rapid exchange occurring at the surface of particles and much slower exchange occurring internally [94]. Short-term equilibrium in the range of several days and long-term equilibrium over several months to several years are assumed to model these exchanges in soils. During periods of rapid transport, the absorption of phosphorus into soil materials decreases due to short contact times, low water-to-soil ratios in the flow path, and possibly facilitated colloidal transport. As a result, heavy rainfall and high concentrations of phosphorus may have strong impacts on the dynamics of phosphorus in agricultural basins with low drainage, and high concentrations of phosphorus may occur during periods of high flow. However, true equilibrium is rarely achieved.

Isotherm equations are used to study the mechanisms of phosphorus absorption in sediments and soil [84]. These equations allow for the determination of important absorption parameters, such as the phosphorus concentration at the equilibrium point ($EPC_0$), the degree of phosphorus saturation, and absorption energy [86, 95, 96]. If the $EPC_0$ and degree of phosphorus saturation in the soil are low and absorption energy is high, the sediments will have a high capacity to absorb phosphorus. Comparing the $EPC_0$ and the concentration of reactive phosphorus in the water solution (DRP) can determine whether the sediments act as a phosphorus reservoir or a source of phosphorus.

In general, there are three main reasons for using absorption isotherms: (1) identification of compounds that play a role in the absorption and release of phosphorus, (2) forecasting the amount of fertilizer needed to maximize production, and (3) studying the nature of absorption to understand the mechanism of these processes [87].

To study phosphorus absorption in sediments, a surface adsorption isotherm is generated by shaking the sediment sample with increasing concentrations of the ion solution. The amount of ion absorbed in each sample is calculated by measuring the difference between the initial concentration and the equilibrium concentration. Laboratory methods, such as those developed by Nelson and Logan in 1983, are commonly used to measure the characteristics of phosphorus absorption in sediments. The results obtained from these methods are then fitted with standard absorption isotherm models, such as Langmuir, Freundlich, and Temkin, as described in the studies by Rhue and Harris [97] and Graetz and Nair [98]. These equations are suitable for methods that reach equilibrium in a short time, but their range of success is limited to a certain concentration of phosphorus, as reported by [99].

Various equations have been employed to elucidate the relationship between the amount of phosphorus absorbed per unit weight of the adsorbent and the concentration of phosphorus in the solution.

### *4.2.1 Langmuir equation*

The Langmuir equation is widely used in soil science. It was first used by Fried and Shapiro [100] and later by Olsen and Watanabe [99] to explain phosphate absorption in soil. The Langmuir equation is based on three main assumptions [101]:

1. The surface absorption energy is constant and independent of the amount of surface coverage, which means the surface is homogeneous.
2. Surface absorption occurs at specific sites without any interaction between the absorbing molecules.
3. The maximum surface absorption corresponds to a complete molecular layer covering the entire active surface of the absorber.

The general form of the Langmuir equation is as follows (Eq. 2):

$$S = \frac{bKC}{1+KC} \tag{2}$$

The surface adsorption isotherm can be expressed using the Langmuir equation, where S represents the amount of absorption, b (also known as $S_{max}$) is the maximum absorption of a single layer that occurs as the equilibrium concentration, C, increases, and K is a parameter that reflects the surface's absorption capacity. The Langmuir equation has one main advantage: it reveals the maximum absorption capacity of a surface ($S_{max}$). Although other equations, such as Redlich-Peterson and Fowler-Guggenheim, can predict the maximum absorption, they are more complex and challenging to use. In comparison to other isotherm equations, the Langmuir equation offers more comprehensive information about phosphorus absorption in soil and sediment [102]. Haines et al. [103] has utilized mathematical methods to further explore the features of these equations and showed that the slope of the Langmuir equation at a concentration close to zero is equal to:

$$\lim_{c \to 0} \frac{ds}{dc} = bK \tag{3}$$

She also showed that we will achieve maximum absorption on surfaces when the equilibrium concentration is large enough:

$$\lim_{c \to \infty} s = \lim_{c \to \infty} \frac{bKC}{1+KC} = \frac{bK}{K} = b \tag{4}$$

In this equation, it is assumed that the absorbing sites do not participate in the absorption after absorbing phosphorus. In simpler terms, phosphorus absorption on absorbent surfaces occurs as a single layer [97, 102]. The graphic form of this equation is in **Figure 6**.

$S_o$: It is known as the amount of primary phosphorus in the absorption phase. $EPC_o$ is also known as the concentration of phosphorus at the equilibrium point. At this concentration, the absorption and excretion of phosphorus in sediments reach equilibrium [93, 104]. Between So and $EPC_o$, phosphorus is released into the water. The slope of the isotherm line, referred to as K or absorption energy as mentioned before, represents the rate of change. In this graph, it is observed that after the initial rapid absorption, the amount of absorption gradually increases with the equilibrium concentration until reaching the maximum absorption ($S_{max}$). After reaching $S_{max}$, the slope of the line remains constant.

### *4.2.2 Freundlich equation*

The Freundlich equation is commonly used for describing phosphorus absorption in soils and has the ability to fit absorption data quite well in most soil types [101]. Freundlich discovered a certain relationship that describes absorption from dilute solutions. This equation is primarily derived from experimental observations, but it can also be obtained theoretically by assuming that the bond energy decreases exponentially with the surface coverage, which is likely closer to the actual absorption conditions [105].

The Freundlich equation is considered superior to the Langmuir equation because it is simple and based on more realistic assumptions. It can account for non-ideal absorption on heterogeneous surfaces and the absorption of multiple layers. Assuming that the reduction in absorption energy with increasing surface coverage is due to the uniformity of the absorbing surfaces justifies the Freundlich equation [101]. The Freundlich equation is commonly defined as Eq. 5:

$$S = K_f C^{\alpha} \tag{5}$$

In this equation, α (alpha) and $K_f$ are called adjustable positive values, where α is usually between zero and one, but in some cases it is seen that it is more than one. By using logarithms on equal sides, this equation becomes a straight line. Although the

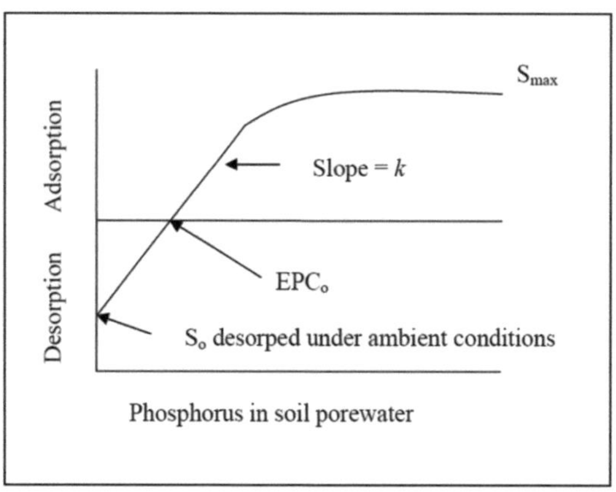


**Figure 6.**
*Schematic display of Longmuir isotherm.*

equation itself is sometimes used unchanged for simplicity. This equation with all its ability is not able to predict the maximum amount of absorption, that is, if the equilibrium concentration moves toward its maximum possible value, the amount of absorption also moves toward its maximum value, but it does not reach a constant value. (Opposite to the Langmuir equation), that is:

$$\lim_{C \to \infty} S = \lim_{C \to \infty} K_f C = K_f(\infty) = \infty \quad (6)$$

Therefore, this equation suggests that as the equilibrium concentration increases, the absorption on the surfaces also increases, and this increase in the amount of absorption is exponential according to the equation. According to some researchers, this equation is better at low equilibrium concentrations [2].

#### *4.2.3 Tamkin's equation*

This equation has found a special place in soil science due to its simpler formula compared to other equations.

$$S = K_1 \; ln \; c + K_2 \quad (7)$$

In this equation, $K_1$ and $K_2$ are constant values. The weak point of this equation is the inability to predict the maximum amount of absorption on the surface [103].

#### *4.2.4 Risk assessment indicators*

In order to identify the points of the sediments that have the greatest potential to release phosphorus into the water, two important parameters, the equilibrium concentration of phosphorus at the zero point ($EPC_o$) and the degree of phosphorus saturation (DPS) are used.

##### *4.2.4.1 Phosphorus concentration at the equilibrium point*

The most important result obtained from isotherm studies is the equilibrium phosphorus concentration ($EPC_o$), which represents the equilibrium between absorption and desorption processes in pollution studies [36, 95]. By comparing the $EPC_o$ value to the concentration of dissolved reactive phosphorus (DRP) in the water solution, it is possible to determine whether sediments act as a source or sink of phosphorus. If $EPC_o$ is higher than the DRP concentration ($EPC_o > DRP$), phosphorus will be released from the sediments into the water column. Conversely, if $EPC_o$ is lower than the DRP concentration ($EPC_o < DRP$), sediments are absorbing phosphorus from the water column [84]. In simpler terms, sediments can either release or absorb phosphorus depending on the $EPC_o$ value, with higher $EPC_o$ values indicating a higher likelihood of phosphorus being released from sediments into the water column. The $EPC_o$ value is graphically represented by the point where the isotherm curve intersects the x-axis (equilibrium concentration axis).

##### *4.2.4.2 Degree of phosphorus saturation in sediments*

The investigation of the degree of phosphorus saturation can be traced back to the research of Breeuwsma and Silva [43], who aimed to establish a correlation between soil phosphorus content and its transfer to groundwater in Dutch soil. Since then, many researchers have used and modified their approach [86, 106, 107] to determine the critical level of phosphorus saturation. Points with a degree of saturation higher than this critical limit are more likely to release phosphorus into water compared to those with lower saturation levels. Breeuwsma and Silva [43] determined the critical level of phosphorus saturation in Dutch soils to be 25%. Sallade and Sims [40] raised this limit to 40% for sediments in American drains. Ngoyan and Sukias [86] found that although the degree of phosphorus saturation in some sediments located in drains was between 64 and 68%, the dissolved phosphorus concentration in the water of these drains was still low, indicating that the level was not sufficient to cause enrichment. On average, the degree of phosphorus saturation increased to more than 65%. The degree of phosphorus saturation reflects the actual state of phosphorus in the soil [108]. It indicates the amount of phosphorus accumulation in sediments relative to the maximum absorption capacity of phosphorus [86], as expressed in Eq. 8:

$$DPS(\%) = \frac{TP}{PSC} \times 100 \tag{8}$$

That DPS is the degree of phosphorus saturation in percentage and PSC indicates the maximum absorption capacity of phosphorus at the desired depth from the soil [17]. TP is the amount of total phosphorus (which is obtained by digesting the sediment sample). The higher the degree of saturation in sediments, the greater the risk of releasing phosphorus into water [86]. Various methods have been used to measure the degree of saturation. Sometimes formula 9 is used in acid soils:

$$DPS(\%) = \frac{P_{ox}}{Fe_{ox} + Al_{ox}} \times 100 \tag{9}$$

$P_{ox}$ in this formula is phosphorus extractable by ammonium oxalate and $Fe_{ox}$ + $Al_{ox}$ is the sum of iron and aluminum extractable by ammonium oxalate [109, 110]. Sometimes Olsen or Mehlich phosphorus together with single point absorption index is used to determine the degree of saturation [102, 111]. These methods are referred to as indirect methods. From the direct method to measure the degree of phosphorus saturation, we can refer to the isotherm equations. Which is obtained by determining the amount of absorbed phosphorus and total phosphorus in sediments.

Methods such as the first two mentioned can be used to determine the degree of phosphorus saturation in specific soils, but the isotherm method is also time-consuming [108]. To address these challenges, Pöthig et al. [108] proposed a simpler method. They demonstrated that this method is applicable to all soil types and is independent of land use.

Water-soluble phosphorus (WSP) is determined based on the ratio of phosphorus absorption to total phosphorus (SP/TP). Each soil has its own level of phosphorus absorption and total phosphorus, but dividing these values cancels out their mutual effects:

$$DPS(\%) = \frac{TP}{PSC} \times 100 \tag{10}$$

$$PSC = TP + SP \tag{11}$$

where SP is known as absorbed phosphorus and is calculated by isotherm experiments [108]. From the combination of the above two equations, the following relationship is obtained:

$$DPS(\%) = \frac{1}{1 + \frac{SP}{TP}} \times 100 \tag{12}$$

Therefore, WSP can be used instead of SP/TP ratio:

$$DPS(\%) = \frac{1}{1 + f(WSP)} \times 100 \tag{13}$$

After examining more than 400 different soil samples (sandy, loamy, peaty and calcareous soils), the following relationship was proposed:

$$DPS(\%) = \frac{1}{1 + 1.25 \times WSP^{-0.75}} \times 100 \tag{14}$$

In this method, when the concentration of phosphorus soluble in water exceeds 5 mg/kg, the degree of phosphorus saturation approaches 70–80%, which indicates a high risk of phosphorus transfer from soil to water. Due to the ease of determining the degree of phosphorus saturation, this method can determine the dangerous points easily and in the shortest time [108].

## 5. Conclusion

Phosphorus is a chemical element with the symbol "P" and atomic number 15. It is a fundamental element for life, playing a vital role in biological processes, energy transfer, and genetic information storage. While it is crucial for ecosystems, managing its levels in the soil-water-sediment environment is essential to prevent pollution therein and maintain their health. Notably, the transfer of phosphorus from soil to water is a complex and significant process that plays a crucial role in shaping the nutrient dynamics and ecological health of soil-water-sediment environment. The aspects discussed in this chapter may help in understanding the major insights into this transfer, which in turn may help in the effective management for preventing phosphorus pollution and maintaining soil-water-sediment health and quality. Further, a sound knowledge of the nutrient enrichment processes; pathways of phosphorus entry; sediment transport; ecological consequences; major ecosystem disturbances; mitigation strategies; regulatory actions; research and monitoring; and collaborative efforts is required for getting more insights into the implications and importance of phosphorus transfer processes in the soil-water-sediment environment.

## Author details

Eisa Ebrahimi[1]* and Mohammad Reza Ojani[2]

1 Soil Science Department, University of Guilan, Rasht, Iran

2 Faculty of Science and Technology, Physical Geography Geoinformatics Department, University of Debrecen, Hungary

*Address all correspondence to: ebrahimi.soilphysic@yahoo.com

## References


[1] Holford ICR. Evalution of soil phosphorus buffering indices. Soil Research. 1979;**17**:495

[2] Tisdal SL, Nelson WL, Beaton JD. Soil Fertility and Fertilizers. 4th ed. New York: Macmillon Publishing Compan; 1984

[3] Ibrahim M, Iqbal M, Tang YT, Khan S, Guan DX, Li G. Phosphorus mobilization in plant–soil environments and inspired strategies for managing phosphorus: A review. Agronomy. 2022; **12**(10):2539

[4] Luo D, Wang L, Nan H, Cao Y, Wang H, Kumar TV, et al. Phosphorus adsorption by functionalized biochar: A review. Environmental Chemistry Letters. 2023;**21**(1):497-524

[5] Hanley PK, Murphy MD. Soil and fertilizer phosphorus in the Irish ecosystem. Water Research. 1973;7:197-210

[6] Johan PD, Ahmed OH, Omar L, Hasbullah NA. Phosphorus transformation in soils following co-application of charcoal and wood ash. Agronomy. 2021;**11**(10):2010

[7] Parfitt RL. Anion adsorption by soils and soil materials. Advances in Agronomy. 1979;**30**:1-50

[8] Dzombak DA, Gorman MR. Expanding perspectives of element cycling from 1970 to 2020: The influence of Stumm and Morgan. Environmental Science & Technology. 2021;**55**:14342-14346. DOI: 10.1021/acs.est.1c01983

[9] Dixon JB. Kaolin and Serpentine Group Minerals. 2018:467-525. DOI: 10.2136/sssabookser1.2ed.c10

[10] Lindsay WL. Chemical Equilibria in Soil. New York, NY, USA: John Wiley & Sons; 1979

[11] Morgan MA. The behaviour of soil and fertilizer phosphorus. In: Tunney H, Carton OT, Brookes PC, Johnston AE, editors. Phosphorus Loss from Soil to Water. New York, NY, USA: CAB International; 1977. pp. 137-149

[12] Rupngam T, Messiga AJ, Karam A. Solubility of soil phosphorus in extended waterlogged conditions: An incubation study. Heliyon. 2023;**9**(2)

[13] Khan A, Lu G, Ayaz M, Zhang H, Wang R, Lv F, et al. Phosphorus efficiency, soil phosphorus dynamics and critical phosphorus level under long-term fertilization for single and double cropping systems. Agriculture, Ecosystems & Environment. 2018;**256**:1-11

[14] Liu L, Gao Z, Yang Y, Gao Y, Mahmood M, Jiao H, et al. Long-term high-P fertilizer input shifts soil P cycle genes and microorganism communities in dryland wheat production systems. Agriculture, Ecosystems & Environment. 2023;**342**:108226

[15] Delgado A, Scalenghe R. Aspects of phosphorus transfer from soils in Europe. Review article. Journal of Plant Nutrition and Soil Science. 2008;**171**:552-575

[16] Gransee A, Merbach W. Phosphorus dynamics in a longterm P fertilization trial on Luvic Phaeozem at Halle. Journal of Plant Nutrition and Soil Science. 2000;**163**:353-357

[17] Kovar JL, Pierzynski GM. Methods of Phosphorus Analysis for Soils, Sediments, Residuals, and Waters. 2nd ed. Southern Cooperative Series Bulletin, 408. 2009

[18] Rode M, Tittel J, Reinstorf F, Schubert M, Knöller K, Gilfedder B, et al. Seasonal variation and release of soluble reactive phosphorus in an agricultural upland headwater in central Germany. Hydrology and Earth System Sciences. 2023;**27**(6):1261-1277

[19] Ngatia L, Taylor R. Phosphorus eutrophication and mitigation strategies. In: Phosphorus-Recovery and Recycling. 2018. pp. 45-61

[20] Wang Y, Luo D, Xiong Z, Wang Z, Gao M. Changes in rhizosphere phosphorus fractions and phosphate-mineralizing microbial populations in acid soil as influenced by organic acid exudation. Soil and Tillage Research. 2023;**225**:105543

[21] Zheng Y, Sun Y, Zhang Z, Han C, Wang Z, Liu C, et al. Evaluation of the distribution and mobility of labile phosphorus in sediment profiles of lake Nansi, the largest eutrophic freshwater lake in northern China. Chemosphere. 2023:137756

[22] Schindler DW. The evolution of phosphorus limitation in lakes. Science (Washington DC). 1977;**195**:260-262

[23] Ritter W, Shirmohammadi A. Agricultural nonpoint source pollution. In: Watershed Management and Hydrology. 1952. pp. 135-145

[24] Mcdowell RW, Sharpley AN, Folmar G. Modification of phosphorus export from an eastern USA catchment by fluvial sediment and phosphorus inputs. Agriculture, Ecosystems & Environment. 2003;**99**:187-199

[25] Lory JA. Agriculture Phosphorus and Water Quality. Department of Agronomy and Commercial Agriculture Programme. Published by MU Extension, University of Missouri Colombia; 1995

[26] Wood CW, Mullins GL, Hajek BF. Phosphorus in Agriculture. Soil Quality Institute. Technical Pamphlet. No. 2; 2005

[27] Riemersma S, Little J, Ontkean G, Moskal-Hebert T. Phosphorus Sources and Sinks in Watersheds: A Review, Alberta Soil Phosphorus Limits Project. Alberta, Canada: Irrigation Branch Alberta Agriculture, Food and Rural Development Lethbridge; 2006

[28] Torbert HA, Daniel TC, Lemunyon JL, Jones RM. Relationship of soil test phosphorus and sampling depth to runoff phosphorus in calcareous and noncalcareous soils. Journal of Environmental Quality. 2002;**31**:1380-1387. DOI: 10.2134/jeq2002.1380

[29] Keup LE. Phosphorus in flowing waters. Water Research. 1968;**2**(5):373-386

[30] Kronvang B, Laubel A, Grant R. Suspended sediment and particulate phosphorus transport and delivery pathways in an arable catchment, Gelbaek stream, Denmark. Hydrological Processes. 1997;**11**(6):627-642

[31] Sharply AN, Menzel RG. The impact of soli and fertilizer phosphorus on the environment. Advances in Agronomy. 1987;**41**:297-324

[32] Heckrath G, Brookes PC, Poulton PR, Goulding KWT. Phosphorus eaching from soils containing different phosphorus concentrations in the Broadbalk experiment. Journal of Environmental Quality. 1995;**24**:904-910

[33] Heathwaite AL. Sources and pathways of phosphorus loss from agriculture. In: Tunney H, Carton OT,

Brookes PC, Johnston AE, editors. Phosphorus Loss from Soil to Water. New York, NY, USA: CAB International; 1997. pp. 205-223

[34] Nasir Ahmad NSB, Mustafa FB, Didams G. A systematic review of soil erosion control practices on the agricultural land in Asia. International Soil and Water Conservation Research. 2020;**8**(2):103-115

[35] Aziz S, Islam MS. Erosion and runoff reduction potential of vetiver grass for hill slopes: A physical model study. International Journal of Sediment Research. 2023;**38**(1):49-65

[36] Sharpley AN, Smith SJ, Jones OR, Berg WA, Coleman GA. The transport of bioavailable phosphorus in agricultural runoff. Journal of Environmental Quality. 1992;**21**:30-35. DOI: 10.2134/jeq1992.00472425002100010003x

[37] Ebrahimi E, Asadi H, Rahmani M, Farhangi MB, Ashrafzadeh A. Effect of precipitation and sediment concentration on the loss of nitrogen and phosphorus in the Pasikhan River. AQUA—Water Infrastructure, Ecosystems and Society. 2022b;**71**(2): 211-228

[38] Sharpley AN, Daniel TT, Sims J. Agricultural Phosphorus and Eutrophication. 2nd ed. United States Department of Agriculture, ARS–149; 2003

[39] Schoumans OF, Chardon WJ. Risk assessment methodologies for predicting hosphorus losses. Journal of Plant Nutrition and Soil Science. 2003;**166**: 403-408

[40] Sallade YE, Sims JT. Phosphorus transformations in the sediments of Delaware's agricultural drainageways: II. Effect of reducing conditions on phosphorus release. Journal of Environmental Quality. 1997;**26**:1579-1588. DOI: 10.2134/jeq1997.00472425002600060018x

[41] Withers PJA. The Significance of Agricultural as a Source of Phosphorus Pollution to Inland and Costal Waters in the UK. Winchester, UK: ADAS Bridgets; 1994. p. 57

[42] Holford ICR, Mattingly GEG. A model for the behavior of labile phosphate in soil. Plant and Soil. 1976; **44**:219-229

[43] Breeuwsma A, Silva S. Phosphorus Fertilization and Environmental Effects in the Netherlands and the Po Region (Italy). Report No. 57. Wageningen, The Netherlands: Agricultural Research Department, The Winand Staring Centre for Integrated Land, Soil and Water Rese; 1992

[44] Newman EI. Phosphorus inputs to terrestrial systems. Journal of Ecology. 1995;**83**:713-726

[45] Dethier DP. Weathering rates and chemical flux from catchments in the Pacific Northwest, USA. In: Colman SM, Dethier DP, editors. Rates of Chemical Weathering of Rocks and Minerals. New York: Academic Press; 1986. pp. 503-530

[46] Yang YG, He ZL, Lin Y, Stoffella PJ. Phosphorus availability in sediments from a tidal river receiving runoff water from agricultural fields. Agricultural Water Management. 2010;**97**:1722-1730

[47] Ibrahim M, Labaki M, Giraudon JM, Lamonier JF. Hydroxyapatite, a multifunctional material for air, water and soil pollution control: A review. Journal of Hazardous Materials. 2020; **383**:121139

[48] Jia Y, Sun S, Wang S, Yan X, Qian J, Pan B. Phosphorus in water: A review on the speciation analysis and species specific removal strategies. Critical Reviews in Environmental Science and Technology. 2023;**53**(4):435-456

[49] Boström B, Pettersson K. Different patterns of phosphorus release from lake sediments in laboratory experiments. Hydrobiologia. 1982;**91**:415-429

[50] Chen YSR, Butler JN, Werner S. Kinetic study of phosphate reaction with aluminum oxide and kaolinite. Environmental Science & Technology. 1973;7:327-332. DOI: 10.1021/es60076a007

[51] Yang X, Zhang R, Wang J, He K, Chen J. Fluxes and mechanisms of phosphorus release from sediments in seasonal hypoxic reservoirs: A simulation-based experimental study. Journal of Soils and Sediments. 2021;**21**: 3246-3258

[52] Boström B, Andersen JM, Fleischer S, Jansson M. Exchange of phosphorus across the sediment-water interface. In: Phosphorus in Freshwater Ecosystems: Proceedings of a Symposium; 25–28 September 1985; Uppsala, Sweden. Springer Netherlands; 1988. pp. 229-244

[53] Lehtoranta J. Benthic phosphorus release from sediment to water. In: Wassmann P, Olli K, editors. Drainage Basin Nutrient Inputs and Eutrophication: An Integrated Approach. Troms, Norway: Norwegian College of Fishery Sciences, University of Trosm; 2004. Available from: http://lepo.it.da.ut.ee/~olli/eutr/html/htmlBook95.html

[54] Stumm W, Kummert R, Sigg L. A ligand exchange model for the adsorption of inorganic and organic ligands at hydrous oxide interfaces. Croatica Chemica Acta. 1980;**53**(2):291-312

[55] Ehrlich HL, Newman DK. Geomicrobiology. 5th ed. New York: CRC Press, Taylor & Francis Group; 2008. pp. 219-232

[56] Gatcher RJ, Meyer JS, Mares A. Contribution of bacteria to release and fixation of phosphorus in lake sediments. Limnology and Oceanography. 1988;**33**: 1542-1558

[57] Cooke GD, Welch EB, Peterson SA, Newroth PR. Restoration and Management of Lakes and Reservoirs. 2nd ed. Boca Raton, FL, USA: Lewis Publishers; 2016. p. 548

[58] Marsden MW. Lake restoration by reducing external phosphorus loading: The influence of sediment phosphorus release. Freshwater Biology. 1989;**21**:139-162

[59] Nielsen T, Andersen FO. Phosphorus dynamics during decomposition of mangrove (Rhizophora apiculata) leaves in sediments. Journal of Experimental Marine Biology and Ecology. 2003;**293**: 73-88

[60] Rast W, Thornton JA. Trends in eutrophication research and control. Hydrol Process. 1996;**10**:295-313. DOI: 10.1002/(SICI)1099-1085(199602)10:2<295::AID-HYP360>3.0.CO;2-F

[61] Kumar A, Mishra S, Bakshi S, Upadhyay P, Thakur TK. Response of eutrophication and water quality drivers on greenhouse gas emissions in lakes of China: A critical analysis. Ecohydrology. 2023;**16**(1):e2483

[62] Klein G, Perera P. Eutrophication and Health. Luxembourg: Environment Quality and Natural Resources European Commision; 2002

[63] Chorus I, Welker M. Toxic Cyanobacteria in Water. London: CRC Press; 2021. DOI: 10.1201/9781003081449

[64] Nürnberg GK. Trophic state of clear and colored, soft-and hardwater lakes with special consideration of nutrients, anoxia, phytoplankton and fish. The Lake and Reservoir Management Journal. 1996;**12**(4):432-447

[65] Song K, Burgin AJ. Perpetual phosphorus cycling: Eutrophication amplifies biological control on internal phosphorus loading in agricultural reservoirs. Ecosystems. 2017;**20**: 1483-1493

[66] Jargal N, Lee EH, An KG. Monsoon-induced response of algal chlorophyll to trophic state, light availability, and morphometry in 293 temperate reservoirs. Journal of Environmental Management. 2023;**337**:117737

[67] Maloney TE, Miller WE, Shiroyama T. Algal responses to nutrient additions in natural waters. I. Laboratory assays. In: Likens GE, editor. Nutrient and Eutrophication. Allen Press; 1972. pp. 134-140

[68] Droop MR. An approach to quantitative nutrition of the phytoplankyon. The Journal of Protozoology. 1977;**24**:528-531

[69] Wynne D, Rhee GY. Effects of light intensity and quality on the relative N and P requirement (the optimum N/P ratio) of marine planktonic algae. Journal of Plankton Research. 1986;**8**:91-103

[70] Correll DL. The role of phosphorus in the eutrophication of receiving waters: A review. Journal of Environmental Quality. 1998;**27**:261-266

[71] Elser JJ, Matzolf ER, Goldman CR. Phosphorus and nitrogen limitation of phytoplankton growth in the freshwaters of North America: A review and critique of experimental enrichments. Canadian Journal of Fisheries and Aquatic Sciences. 1990;**47**:1468-1477

[72] Ryther JH, Dunstan WM. Nitrogen, phosphorus, and eutrophication in the coastal marine environment. Science (Washington DC). 1971;**171**:1008-1013

[73] Hecky RE, Kilham P. Nutrient limitation of phytoplankton in freshwater and marin environments: A review of recent evidence on the effects of enrichment. Limnology and Oceanography. 1988;**33**:796-822

[74] Sonzongni WC, Chapra SC, Armstrong DE, Logan TJ. Bioavailability of phosphorus inputs to lakes. Journal of Environmental Quality. 1982;**11**:555-563

[75] Boynton WR, Kemp WM. Nutrient regeneration and oxygen consumption by sediments salinity gradient. Marine Ecology Progress Series. 1985;**23**:45-55

[76] Froelich PN. Kinetic control of dissolved phosphorus in natural rivers and estuaries: A primer on the phosphate buffer mechanism. Limnology and Oceanography. 1988;**33**:649-668

[77] Hutchinson GE. A Treasties on Limnology. Vol. 1. New York: John Wiley & Sons; 1957

[78] Wauchope RD, McDowell LL. Adsorptino of phosphate, arsenate, methanearsonate and cacodylate by lake and stream sediments: Comparisons with soils. Journal of Environmental Quality. 1984;**13**:499-504

[79] Forstner U, Salomons W. Trace metal analysis on plluted sediments: Part 2, evaluation of environmental impact.

Environmental Technology. 1980;**11**:506-517

[80] Mcdowell RW, Sharpley AN, Condron LM, Haygrath PM, Brookes PC. Processes controlling soil phosphorus release to runoff and implications for agricultural managements. Nutrient Cycling in Agroecosystem. 2001;**59**:269-284

[81] Singh PK, Bhunya PK, Mishra SK, Chaube UC. A sediment graph model based on SCS-CN method. Journal of Hydrology. 2008;**349**:244-255

[82] Blanco-Canqui H, Lal R. Principles of Soil Conservation and Management. Springer Science; 2008. p. 617

[83] Sposito G. The Chemistry of Soils. New York: Published by Oxford University Press, Inc.; 1989

[84] Wang Y, Shen Z, Niu J, Liu R. Adsorption of phosphorus on sediments from the Three-Gorges Reservior (China) and the reletion with sediment compositions. Hazardous Materials. 2009;**162**:92-98

[85] Arias M, Da Silva-Carballal J, García-Río L, Mejuto J, Núñez A. Retention of phosphorus by iron and aluminum-oxides-coated quartz particles. Journal of Colloid and Interface Science. 2006;**295**:65-70

[86] Neguyen L, Sukias J. Phosphorus fractions and retention in dranage dich sediments receiving surface runoff and subsurface drainage from agricultural catchments in the North Island, New Zealand. Agriculture, Ecosystems & Environment. 2002;**92**:49-69

[87] Sanyal SK, De Datta SK. Chemistry of phosphorus transformations in soil. Advances in Soil Science. 1991;**16**:1-120

[88] Walter LM, Morse JW. Reactive surface area of skeletal carbonates during dissolution: Effect of grain size. Journal of Sedimentary Research. 1984;**54**:1081-1090

[89] Zhang JZ, Huang X. Relative importance of solid-phase phosphorus and iron on the sorption behavior of sediments. Environmental Science and Technology. 2007;**41**:2789-2795

[90] Wang Q, Li Y. Phosphorus adsorption and desorption behavior on sediments of differen origins. Journal of Soils and Sediments. 2010;**10**:1159-1173

[91] Lake BA, Coolidge KM, Norton SA, Amirbahman A. Factors contributing to the internal loading of phosphorus from anoxic sediments in six Maine, USA, lakes. Science of The Total Environment. 2007;**373**:534-541

[92] Nwoke OC, Vanlauwe B, Diels J, Sanginga N, Osonubi O, Merckx R. Assessment of labile phosphorus fractions and adsorption characteristics in relation to soil properties of west African savanna soils. Agriculture, Ecosystems & Environment. 2003;**100**:285-294

[93] Diaz OA, Reddy KR, Moor J. Solublity of inorganic phosphorus in stream water as influenced by pH and calcium concentration. Water Research. 1994;**28**(8):1755-1763

[94] Barrow NJ. The four laws of soil chemistry. Australian Journal of Soil Research. 1999;**37**:787-829

[95] Haggard BE, Stanley EH, Storm DE. Nutrient retention in a point-source enriched stream. Journal of the North American Benthological Society. 2005;**24**:29-47

[96] Yoo JH, Ro HM, Choi WJ, Yoo SH, Han KH. Phosphorus adsorption and removal by sediments of a constructed

marsh in Korea. Ecological Engineering. 2006;**27**:109-117

[97] Rhue RD, Harris RG. Phosphorus sorption/desorption reactions in soils and sediments. In: Reddy KR, O'Connor GA, Schleske CL, editors. Phosphorus Biogeochemistry in Subtropical Ecosystems. Boca Raton: Lewis Publishers; 1999. pp. 187-206

[98] Graetz DA, Nair VD. Phosphorus sorption isotherm determination. In: Pierzynski GM, editor. Methods of Phosphorus Analysis for Soils, Sediments, Residuals and Water. Manhattan: North Carolina State University; 2000. pp. 35-38

[99] Olsen SR, Wanatabe FS. A method to determine phosphorus adsorption maximum of soils as measured by the Langmuier isotherm. Soil Science Society of America Journal. 1957;**21**:144-149

[100] Fried M, Shapiro G. Phosphorus supply pattern of various soils. Soil Science Society of America, Proceedings. 1956;**20**:471-475

[101] Bohn H, McNeal B, Oconner G. Soil Chemistry. 2nd ed. New York: John Wiley & Sons; 1985. p. 329

[102] Nair VD, Portier KM, Graetz DA, Walker ML. An environmental threshold for degree of phosphorus saturation in sandy soils. Journal of Environmental Quality. 2004;**33**:107-113

[103] Haines J, Léger J, Bocquillon G. Synthesis and design of superhard materials. Annual Review of Materials Research. 2001;**31**:1-23. DOI: 10.1146/annurev.matsci.31.1.1

[104] Taylor AW, Kunishi HM. Phosphate equilibria on stream sediment and soil in a watershed draining an agricultural region. Journal of Agricultural and Food Chemistry. 1971; **19**:827-831

[105] Sibbesen E. Some new equations to describe phosphate sorption by soils. Soil Science Society of America Journal. 1981; **32**:67-74

[106] Zhou M, Li Y. Phosphorus sorption characteristics of calcareous soils and limestone from the southern everglades and adjacent farmlands. Soil Science Society of America Journal. 2001;**65**: 1404-1412

[107] Sims JT, Edwards AC, Schoumans OF, Simard RR. Integrating soil phosphorus testing into environmentally based agricultural management practices. Journal of Environmental Quality. 2000;**29**:60-71

[108] Pöthig R, Behrendt H, Opitz D, Furrer G. A universal method to assess the potential of phosphorus loss from soil to aquatic ecosystems. Environmental Science and Pollution Research. 2010;**17**:497-504

[109] Maguire RO, Sims JT. Measuring agronomic and environmental soil phosphorus saturation and predicting phosphorus leaching with Mehlich 3. Soil Science Society of America Journal. 2002;**66**:2033-2039

[110] Börling K, Otabbong E, Barberis E. Soil variables for predicting potential phosphorus release in Swedish noncalcareous soils. Journal of Environmental Quality. 2004;**33**:99-106

[111] Simard RR, Cluis D, Gangbazo G, Beauchemin S. Phosphorus status of forest and agricultural soils from a watershed of high animal density. Journal of Environmental Quality. 1995; **24**:1010-1017

Chapter 3

# Role of Phosphorus in the Photosynthetic Dark Phase Biochemical Pathways

*Alex Odoom and Wilfred Ofosu*

## Abstract

Phosphorus (P) is an essential mineral nutrient for plant growth and development, second only to nitrogen in abundance. It is frequently limited in soil, requiring the application of P-fertilizers to improve plant productivity. One critical function of P in plants is its role in the dark phase of photosynthesis, where it functions in energy storage and transfer, carbon fixation, regulation of the dark phase, and nucleotide and coenzyme biosynthesis. P is a foundational component of important molecules like ATP and essential coenzymes, which are crucial for efficient carbon fixation and energy conversion during the Calvin cycle. Sustainable P-management strategies and improved agricultural practices are necessary to optimize plant growth and ensure sustainable agricultural production in the face of P-limitations.

**Keywords:** Calvin-Benson-Bassham (CBB) cycle, phosphorylation, adenosine triphosphate, nicotinamide adenine dinucleotide phosphate, phosphorus, photophosphorylation

## 1. Introduction

Photosynthesis is the process by which plants, algae, and some bacteria convert light energy into chemical energy in the form of glucose and other organic molecules [1]. This process is essential for the survival of nearly all life on Earth, as it forms the basis of the food chain and is responsible for the production of oxygen in the atmosphere [2]. Photosynthesis consists of two main phases: the light-dependent reactions, which occur in the thylakoid membranes of the chloroplasts and generate ATP and NADPH; and the light-independent reactions, also known as the dark phase or the Calvin-Benson-Bassham (CBB) cycle, which occur in the stroma of the chloroplasts and utilize the ATP and NADPH produced in the light-dependent reactions to fix CO2 into organic molecules [3].

Phosphorus (P) is an essential element for all living organisms, as it is a key component of several biomolecules, such as nucleic acids, ATP, and phospholipids. P is beneficial to plants both in its elemental form and as a component of other molecules such as phosphates. After nitrogen (N), P is quantitatively the most significant inorganic nutrient for plant growth, and often limits primary productivity in natural systems as well as agricultural systems, unless supplied as fertilizer [4]. In plants, P plays a crucial role in various physiological processes, including energy metabolism, nucleic acid synthesis, and membrane function, as well as in the regulation of enzyme activity and signal transduction pathways [5, 6]. P is known to play a critical role in several biochemical pathways involved in the dark phase of photosynthesis, including its involvement in ATP synthesis, NADPH production, and its presence in several key enzymes. Carbon fixation, the initial step in the dark phase of photosynthesis, is a phosphate-driven metabolic process [7].

To minimize the adverse effects of global warming, it is crucial to achieve efficient carbon fixation by increasing vegetation cover and nutrient availability. Baslam et al. highlighted the significant contribution of P in addressing this pressing global challenge, as it impacts the dark phase of photosynthesis directly and indirectly [8]. As global P reserves continue to deplete, it becomes increasingly vital to comprehend the multifaceted role of phosphorus in the dark phase of photosynthesis [9]. In agricultural production, P-deficiency presents a significant limitation, and a better understanding of its contribution to the dark phase of photosynthesis could hold significant implications for improving crop productivity and sustainability [10].

Given above, the goal of this literature review is to provide a detailed understanding of the importance of P in the dark phase of photosynthesis, particularly in the regulation of the CBB cycle and the production of energy-rich molecules, such as ATP and NADPH and its overall significance in the process of photosynthesis.

## 2. Phosphorus in plant growth and development

Plant growth heavily relies on inorganic nutrients; predominantly N, and P both of which are often limited in the environment, hence these nutrients are frequently administered as fertilizers to plants [11]. An optimal P fertilization can enhance plant biomass yield and improve antioxidant potential [12]. Increasing P-levels can also result in improved growth parameters, including fresh and dry weight, floral number, plant height, and essential oil concentration [13]. Additionally, P plays a crucial role as a metabolic and regulatory nutrient element. However, the interaction of P with other minerals is often stronger than its individual action [14]. Particularly in the dark phase of photosynthesis, the availability of P can significantly impact the efficiency of the CBB cycle and the production of energy-rich molecules such as ATP and NADPH.

P plays a crucial role in enhancing a plant's ability to resist and tolerate diseases, which can cause significant reductions in both crop yield and quality [15]. Its application is known to enhance plant growth and also to increase resistance to various biotic stresses. For instance, when a pathogen invades a plant, P in the form of ATP is released into the extracellular space, where the plants recognize it as a signal that there is cellular damage [16]. In turn, this signal activates the plant's defense response to fight off the pathogen. Therefore, ATP is a signaling molecule for the defense response activation in the plant [17]. Further, P-deficiency can have a profound impact on plant growth and productivity, leading to reduced root growth, delayed flowering, impaired seed development, and decreased crop yield [4, 5]. In addition to these direct effects, phosphorus deficiency can also influence the plant's ability to acquire other essential nutrients, such as N, K, and micronutrients, further exacerbate the impact on plant growth and development [18, 19].

## 3. Calvin-Benson cycle: An overview

The dark phase of photosynthesis, or the Calvin-Benson cycle, is a series of enzyme-catalyzed reactions that occur in the stroma of chloroplasts [20]. This cycle is responsible for the fixation of $CO_2$ into oces, ultimately producing glucose and other sugars that can be used for various cellular processes [21]. Calvin-Benson cycle consists of three main stages: carboxylation, reduction, and regeneration [22] (**Figure 1**).

The Calvin cycle, also known as the dark phase of photosynthesis, is a series of enzyme-driven reactions that convert carbon dioxide ($CO_2$) into sugars and other organic molecules. The cycle consists of three main phases: carbon fixation, reduction, and regeneration of the $CO_2$ acceptor molecule ribulose-1,5-bisphosphate (RuBP). During these reactions, P plays a vital role in the form of energy transfer molecules (ATP and NADPH) [23].

### 3.1 Carboxylation

Carboxylation is the first step of the Calvin-Benson cycle, where $CO_2$ is fixed into an organic molecule through the enzyme ribulose-1,5-bisphosphate carboxylase/oxygenase (RuBisCO) [24]. RuBisCO catalyzes the reaction between $CO_2$ and the five-carbon sugar ribulose-1,5-bisphosphate (RuBP), resulting in two molecules of 3-phosphoglycerate (3-PGA) [25]. The carbon fixation phase begins with the enzyme ribulose-1,5-bisphosphate carboxylase/oxygenase (RuBisCO), which catalyzes the

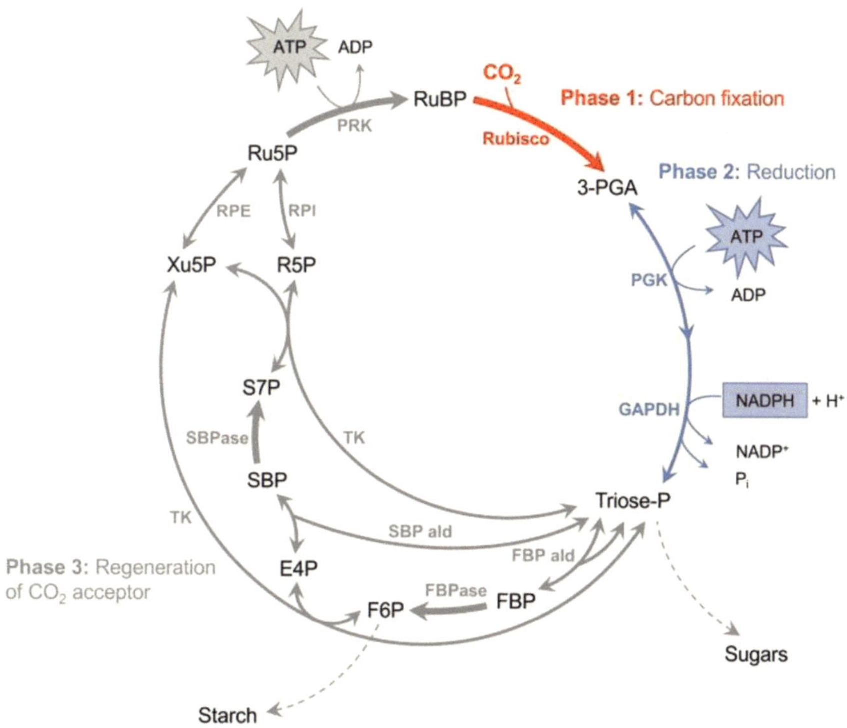


**Figure 1.**
*Calvin–Benson cycle consists of carbon fixation with carbon dioxide, reduction with NADPH and regeneration of the CO2 acceptor. Image adopted from Schreier and Hibberd [22].*

fixation of $CO_2$ to RuBP, producing two molecules of 3-phosphoglycerate (3-PGA) [25]. P is crucial in this step as 3-PGA contains a phosphate group, which is essential for the subsequent steps of the Calvin cycle.

### 3.2 Reduction

The reduction stage involves the conversion of 3-phosphoglycerate (3-PGA) into glyceraldehyde-3-phosphate (G3P) through two enzyme-catalyzed reactions [26, 27]. First, 3-PGA is phosphorylated by the enzyme phosphoglycerate kinase (PGK), yielding 1,3-bisphosphoglycerate (1,3-BPG) [28]. This reaction requires ATP as a substrate, which is converted into ADP during the process. This step highlights the importance of phosphorus, as ATP is the primary energy currency of the cell and contains a high-energy phosphate bond. Next, glyceraldehyde-3-phosphate dehydrogenase (GAPDH) catalyzes the reduction of 1,3-BPG to G3P, using NADPH as a reducing agent [29]. NADPH, which is another essential molecule containing phosphorus, is generated during the light-dependent reactions of photosynthesis.

### 3.3 Regeneration

The final stage of the Calvin-Benson cycle is the regeneration of RuBP from G3P [30]. This process involves a series of enzyme-catalyzed reactions that convert the remaining G3P molecules into RuBP. Enzymes involved in this stage include triose phosphate isomerase (TPI), aldolase, fructose-1,6-bisphosphatase (FBPase), transketolase (TK), and sedoheptulose-1,7-bisphosphatase (SBPase) [31]. Transketolase and aldolase are responsible for the interconversion of carbon skeletons between different sugar phosphate molecules, while phosphoribulokinase catalyzes the phosphorylation of ribulose-5-phosphate (Ru5P) to RuBP, utilizing ATP as a phosphate donor. Once again, phosphorus plays a critical role in the form of ATP, providing the energy required for RuBP regeneration. The regenerated RuBP can then be used for another round of carboxylation, continuing the cycle [32].

## 4. Phosphorus in ATP synthesis and NADPH production

In the dark phase of photosynthesis, P is mainly involved in the synthesis of ATP and the production of NADPH. Both ATP and NADPH are essential energy carriers in the Calvin-Benson cycle, providing the energy and reducing power required for the fixation of $CO_2$ into organic molecules [1].

### 4.1 ATP synthesis

Adenosine triphosphate (ATP) is a high-energy molecule that serves as the primary energy currency for cells [33]. It is synthesized in the light phase of photosynthesis through a process called photophosphorylation, which occurs in the thylakoid membrane of chloroplasts (**Figure 2**). During this process, the energy derived from absorbed light is used to pump protons across the thylakoid membrane, creating a proton gradient [35]. The resulting proton motive force drives the synthesis of ATP from ADP and inorganic phosphate (Pi) through the enzyme ATP synthase [36]. P is a critical component of ATP, as it forms the phosphate groups that store and release energy during ATP hydrolysis. High-energy phosphate, held as a part of the chemical

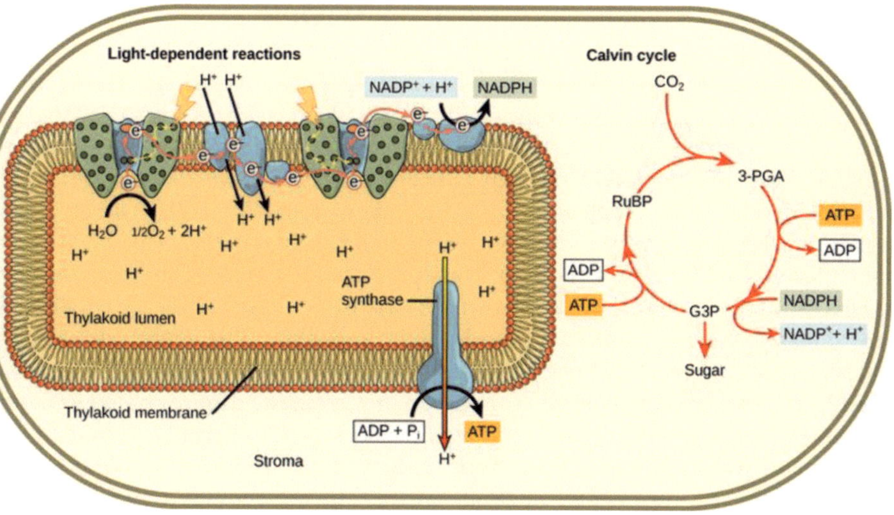


**Figure 2.**
*Light reactions harness energy from the sun to produce chemical bonds, ATP, and NADPH. Image adopted from [34].*

structures of adenosine diphosphate (ADP) and ATP, is the source of energy that drives the multitude of chemical reactions within the plant [33]. When ADP and ATP transfer the high-energy phosphate to other molecules (termed phosphorylation), the stage is set for many essential processes to occur.

Several enzymes are involved in the process of ATP synthesis including ATP synthase [37], which according to several studies is sited in the thylakoid membrane of chloroplasts [38–40]. The activity of ATP synthase, the enzyme responsible for ATP synthesis, is regulated by several factors including the availability of phosphorus [41, 42]. Studies have revealed that P-availability can affect ATP synthesis and cellular energy metabolism. P-deficiency in plants can decrease ATP synthesis and lead to reduced growth and development [9]. This is because P-deficiency inhibits the catalytic activity of ATP synthase, which catalyzes ATP synthesis. Additionally, P-limitation in marine phytoplankton was reported to reduce ATP synthesis [43]. In the dark phase of photosynthesis, ATP is used as an energy source for various enzymatic reactions, such as the phosphorylation of 3-PGA by PGK [44].

### 4.2 NADPH production

Nicotinamide adenine dinucleotide phosphate (NADPH) is another essential energy carrier in the Calvin-Benson cycle, where it provides reducing power for the conversion of 1,3-BPG to G3P by GAPDH [29]. NADPH is produced during the light phase of photosynthesis in a process called linear electron flow (LEF), which occurs in the thylakoid membrane of chloroplasts [45]. During LEF, electrons are transferred from water molecules to $NADP^+$ through a series of protein complexes and electron carriers, including photosystem II (PSII), the cytochrome b6f complex, photosystem I (PSI), and ferredoxin-$NADP^+$ reductase (FNR) [46]. The reduction of $NADP^+$ to NADPH involves the transfer of two electrons and one proton, with the latter being derived from the hydrolysis of water molecules. P is not directly involved in the production of NADPH, but it is essential for the stability and function of $NADP^+$ and NADPH, as they both contain a phosphate group [47].

## 5. Phosphorus as a component of ATP and NADPH

ATP and NADPH are two energy-rich molecules that play a central role in the dark phase of photosynthesis (**Figure 2**). ATP is often referred to as the "energy currency" of the cell, as it can store and release energy in the form of phosphate bonds. In the CBB cycle, ATP provides the energy required for the fixation of CO2 into organic molecules, such as glucose and other sugars. NADPH is an electron carrier that provides the reducing power needed for the synthesis of these organic molecules. Both ATP and NADPH contain phosphate groups, which are essential for their function as energy carriers. In the case of ATP, the molecule consists of three phosphate groups linked to an adenosine molecule. The energy stored in ATP is released when one of the phosphate groups is removed through a process called hydrolysis, resulting in the formation of adenosine diphosphate (ADP) and inorganic phosphate (Pi). This energy can then be used to drive various cellular processes, including the reactions of the CBB cycle [37]. Similarly, NADPH contains a phosphate group as part of its structure. The molecule is derived from its oxidized form, $NADP^+$, by the addition of two electrons and a proton, resulting in the formation of NADPH and the release of a second proton. The reducing power of NADPH is then used in the CBB cycle to convert the fixed CO2 into organic molecules, such as glucose and other sugars [48].

Given the central role of ATP and NADPH in the dark phase of photosynthesis, the availability of P can have a significant impact on the efficiency of the CBB cycle and, consequently, on plant growth and productivity. Indeed, several studies have reported a strong correlation between P-availability and the rate of photosynthesis, with P-deficiency leading to reduced rates of $CO_2$ assimilation and decreased production of ATP and NADPH [49].

## 6. Phosphorus in key enzymes of the dark phase

Several key enzymes in the Calvin-Benson cycle contain P as a major structural component, highlighting the importance of P in the dark phase of photosynthesis. These enzymes include RuBisCO, PGK, GAPDH, and others involved in the regeneration stage of the cycle [50].

### 6.1 RuBisCO

RuBisCO, the most abundant enzyme on Earth, is responsible for the carboxylation of RuBP [25]. The enzyme contains a large subunit and a small subunit, both of which are encoded by the chloroplast genome [51]. The large subunit contains a conserved lysine residue that forms a Schiff base with the phosphate group of RuBP, facilitating the binding of $CO_2$ [52]. The active site of RuBisCO also contains a tightly bound $Mg^{2+}$ ion, which coordinates with the phosphate groups of RuBP and stabilizes the transition state during carboxylation [53].

### 6.2 PGK and GAPDH

Phosphoribuloseglycerate kinase (PGK) and glyceraldehyde-3-phosphate dehydrogenase (GAPDH) are responsible for the phosphorylation and reduction of 3-PGA, respectively [29]. Both enzymes contain P as part of their active sites, with PGK utilizing a phosphohistidine intermediate in its catalytic mechanism, and GAPDH employing a phosphorylated cysteine residue for the transfer of phosphate groups between substrates [29].

### 6.3 Regeneration enzymes

The regeneration stage of the Calvin-Benson cycle involves several enzymes that are responsible for the recycling of the $CO_2$ acceptor molecule, RuBP, which is essential for the continued fixation of $CO_2$ by the cycle. Some of these enzymes, such as SBPase and FBPase, contain P in their active sites, which play a critical role in their catalytic activity [54]. For example, FBPase uses a phosphohistidine intermediate in its reaction mechanism, while SBPase utilizes a phosphoserine intermediate. Phosphorus is also an integral component of RuBP itself, which is regenerated during this stage of the cycle [27].

## 7. Phosphorus in the regulation of key enzymes involved in the CBB cycle

The Calvin-Benson-Bassham (CBB) cycle consists of a series of enzyme-catalyzed reactions that assimilate $CO_2$ into organic molecules, such as glucose and other sugars. These reactions can be grouped into three main stages: (1) carboxylation, in which $CO_2$ is fixed to a five-carbon molecule called ribulose-1,5-bisphosphate (RuBP) by the enzyme ribulose-1,5-bisphosphate carboxylase/oxygenase (Rubisco); (2) reduction, in which the resulting six-carbon molecule is split into two three-carbon molecules of 3-phosphoglycerate (3-PGA) and subsequently reduced to glyceraldehyde-3-phosphate (G3P) using the energy provided by ATP and NADPH; and (3) regeneration, in which the remaining G3P molecules are used to regenerate RuBP through a series of reactions that also involve the consumption of ATP [55].

P plays a crucial role in the regulation of several key enzymes involved in the CBB cycle, including Rubisco, phosphoribulokinase (PRK), and GAPDH. For example, Rubisco, the most abundant enzyme in the chloroplast and responsible for the carboxylation of RuBP, requires a divalent metal ion, such as $Mg^{2+}$, for its catalytic activity [56]. In the dark, the $Mg^{2+}$ ion is replaced by a proton, leading to the formation of a stable complex between Rubisco and a molecule of RuBP. The addition of inorganic phosphate (Pi) can reverse this inhibition by promoting the release of the proton and the re-activation of the enzyme [57]. Therefore, the activity of RuBisCO is regulated by the phosphorylation and dephosphorylation of its activase, RuBisCO activase [58]. The phosphorylation status of RuBisCO activase is modulated by a protein kinase and a phosphatase, which are sensitive to the ATP/ADP ratio in the chloroplast stroma [58]. This regulation ensures that RuBisCO activity is adjusted according to the energy availability within the cell.

Similarly, PRK, the enzyme responsible for the phosphorylation of ribulose-5-phosphate (Ru5P) to RuBP, is also regulated by the availability of Pi. The enzyme is inhibited by the binding of a molecule of ADP, which competes with the substrate, ATP, for the same binding site. The addition of Pi can relieve this inhibition by promoting the formation of ATP from ADP and Pi, thereby allowing the enzyme to resume its catalytic activity [27, 59]. Finally, GAPDH, the enzyme that catalyzes the reduction of 1,3-bisphosphoglycerate (1,3-BPG) to G3P using NADPH as the reducing agent, is also sensitive to changes in the availability of Pi. The enzyme forms a complex with another enzyme, phosphoribuloseglycerate kinase (PGK), and the two enzymes work together in a coupled reaction to convert 3-PGA to G3P. The formation of this complex is dependent on the presence of Pi, which acts as a stabilizing factor and ensures the efficient transfer of phosphate groups between the two enzymes [60, 61].

## 8. Phosphorus and photosynthetic efficiency

The efficient functioning of the dark phase of photosynthesis is directly impacted by the availability of P, a vital component in the production of energy transfer molecules ATP and NADPH [9, 62]. Without sufficient P, plants cannot produce these molecules, leading to reduced photosynthesis and decreased crop yields. Several studies have shown the importance of P in photosynthesis and crop productivity. For example, P-deficiency can significantly reduce the photosynthetic rate and yield [63]. Similarly, P-deficient soils can negatively impact the growth and yield of teff, a staple crop in Ethiopia [64]. Moreover, P-deficiency can also lead to a decrease in the activity of key enzymes involved in the Calvin cycle, such as RuBisCO, further contributing to the reduction in photosynthetic efficiency [65]. Being a key component of ADP and ATP synthase P is required for optimal photosynthetic activity. Hence, P-deficiency limits or halts the biochemical process of phosphorylation and limits the catalytic activity of energy-producing enzymes [66, 67].

Unfortunately, P is often limited in soils, making it a scarce resource for plants. Excessive use of P-fertilizers can also lead to environmental problems. The over-application of P-fertilizers, surpassing crop demand, poses a heightened risk of P-loss from soil to water resources and can result in the degradation of water quality through eutrophication [68]. Therefore, it is important for farmers to understand how to manage phosphorus effectively in their soils to ensure sustainable crop production. One approach to managing P effectively is using precision agriculture techniques. For example, a study conducted in Ghana showed that precision application of P-fertilizer based on soil testing and yield potential resulted in higher crop yields and reduced phosphorus runoff compared to traditional broadcasting of fertilizer [69]. In response to P-deficiency, plants can develop adaptive mechanisms, such as increasing the expression of high-affinity phosphate transporters and enhancing the secretion of acid phosphatases, which help to increase phosphorus uptake and utilization [70]. Plants also modify their metabolic pathways and root morphology, and this involves changes in their gene expression. A range of proteins involved in the various metabolic pathways are differentially expressed in response to phosphate stress, suggesting that they may play important roles in regulating complex adaptation activities for Pi deprivation to facilitate P-homeostasis [71]. The proteins that respond to P shortage may be involved in various processes including phytohormone biosynthesis, signal transduction, cellular organization and defense, and energy and carbon metabolism [72]. These proteins are likely to play a crucial role in sensing the changes in external Pi concentration and regulating complex adaptation activities that enable plants to maintain P homeostasis.

## 9. Conclusions and future perspective

P has been widely known to play a crucial role in the dark phase of photosynthesis, particularly in the Calvin cycle. It is involved in the formation of essential molecules like ATP, NADPH, and sugar phosphates, which are indispensable for the carbon fixation, reduction, and regeneration of RuBP. Additionally, phosphorus is involved in the regulation of key enzymes, such as RuBisCO, through the modulation of its activase. P-availability can directly impact the efficiency of the dark phase of photosynthesis, with P- deficiency leading to reduced photosynthetic rates and adaptive responses in plants. Understanding the role of P in the dark phase of photosynthesis is vital for improving crop productivity and developing sustainable agricultural practices. Future research should focus on identifying novel strategies to enhance P-use efficiency in plants, as well as exploring the potential interactions between phosphorus and other nutrients, such as N and K, in the regulation of photosynthetic processes.

## Acknowledgements

The authors would like to express gratitude to all the national service personnel of the 2022 group at the Department of Medical Microbiology, University of Ghana Medical School, for their invaluable assistance and contributions to this project.

## Conflict of interest

The authors declare no conflict of interest.

## Appendices and nomenclature

| | |
|---|---|
| 3-PGA | 3 -phosphoglycerate |
| ATP | Adenosine triphosphate |
| NADPH | Nicotinamide adenine dinucleotide phosphate |
| Pi | inorganic phosphate |
| P | organic phosphate |
| RuBisCO | Ribulose-1, 5-bisphosphate carboxylase-oxygenase |
| RuBP | ribulose-1-5-bisphosphate |
| CBB | Calvin-Benson-Bassham |

## Author details

Alex Odoom[1*] and Wilfred Ofosu[2,3]

1 Department of Biochemistry, Cell and Molecular Biology, University of Ghana, Legon, Ghana

2 Department of Plant and Environmental Biology, University of Ghana, Legon, Ghana

3 Department of Animal Biology and Conversation Science, University of Ghana, Legon, Ghana

*Address all correspondence to: alexodoom742@gmail.com

## References

[1] Johnson MP. Photosynthesis. Essays Biochemical. 2016;**60**(3):255-273 Available from: https://pubmed.ncbi.nlm.nih.gov/27784776

[2] Janssen PJD, Lambreva MD, Plumeré N, Bartolucci C, Antonacci A, Buonasera K, et al. Photosynthesis at the forefront of a sustainable life. Frontier in Chemistry. 2014;**2**:36. Available from: https://pubmed.ncbi.nlm.nih.gov/24971306

[3] Trinh MDL, Masuda S. Chloroplast pH homeostasis for the regulation of photosynthesis. Frontier in Plant Science. 2022;**13**:919896. Available from: https://pubmed.ncbi.nlm.nih.gov/35693183

[4] Vance CP, Uhde-Stone C, Allan DL. Phosphorus acquisition and use: Critical adaptations by plants for securing a nonrenewable resource. New Phytologist. 2003;**157**(3):423-447. DOI: 10.1046/j.1469-8137.2003.00695.x

[5] Veneklaas EJ, Lambers H, Bragg J, Finnegan PM, Lovelock CE, Plaxton WC, et al. Opportunities for improving phosphorus-use efficiency in crop plants. New Phytologist. 2012;**195**(2):306-320. DOI: 10.1111/j.1469-8137.2012.04190.x

[6] Raghothama KG. Phosphate acquisition. Annual Review in Plant Physiology Plant Molecular Biology. 1999;**50**(1):665-693. DOI: 10.1146/annurev.arplant.50.1.665

[7] Stirbet A, Lazár D, Guo Y, Govindjee G. Photosynthesis: Basics, history and modelling. Annals of Botany. 2020;**126**(4):511-537. Available from: https://pubmed.ncbi.nlm.nih.gov/31641747

[8] Baslam M, Mitsui T, Hodges M, Priesack E, Herritt MT, Aranjuelo I, et al. Photosynthesis in a changing global climate: Scaling up and scaling down in crops. Frontier in Plant Science. 2020;**11**:882. Available from: https://pubmed.ncbi.nlm.nih.gov/32733499

[9] Carstensen A, Herdean A, Schmidt SB, Sharma A, Spetea C, Pribil M, et al. The impacts of phosphorus deficiency on the photosynthetic electron transport chain. Plant Physiology. 2018;**177**(1):271-284 Available from: https://pubmed.ncbi.nlm.nih.gov/29540590

[10] Bechtaoui N, Rabiu MK, Raklami A, Oufdou K, Hafidi M, Jemo M. Phosphate-dependent regulation of growth and stresses management in plants. Frontier in Plant Science. 2021;**12**:679916. Available from: https://pubmed.ncbi.nlm.nih.gov/34777404

[11] Guignard MS, Leitch AR, Acquisti C, Eizaguirre C, Elser JJ, Hessen DO, et al. Impacts of nitrogen and phosphorus: From genomes to natural ecosystems and agriculture. Frontier in Ecology Evolution. 2017;**2017**:5. DOI: 10.3389/fevo.2017.00070

[12] Nell M, Vötsch M, Vierheilig H, Steinkellner S, Zitterl-Eglseer K, Franz C, et al. Effect of phosphorus uptake on growth and secondary metabolites of garden sage (*Salvia officinalis* L.). Journal of Science Food Agriculture. 2009;**89**(6):1090-1096. DOI: 10.1002/jsfa.3561

[13] Chrysargyris A, Panayiotou C, Tzortzakis N. Nitrogen and phosphorus levels affected plant growth, essential oil composition and antioxidant status of lavender plant (*Lavandula angustifolia* Mill.). Industrial Crops Products. 2016;**83**:577-586. DOI: 10.1016/j.indcrop.2015.12.067

[14] Wu W, Zhu S, Chen Q, Lin Y, Tian J, Liang C. Cell Wall proteins play critical roles in plant adaptation to phosphorus deficiency. International Journal of Molecular Science. 2009;**20**(21):5259 Available from: https://pubmed.ncbi.nlm.nih.gov/31652783

[15] Ainsworth EA, Long SP. 30 years of free-air carbon dioxide enrichment (FACE): What have we learned about future crop productivity and its potential for adaptation? Global Change Biology. 2020;**27**(1):27-49. DOI: 10.1111/gcb.15375

[16] Kumar S, Tripathi D, Okubara PA, Tanaka K. Purinoceptor P2K1/DORN1 enhances plant resistance against a Soilborne fungal pathogen, *Rhizoctonia solani*. Frontier in Plant Science. 2020;**11**:572920. Available from: https://pubmed.ncbi.nlm.nih.gov/33101341

[17] Tripathi D, Zhang T, Koo AJ, Stacey G, Tanaka K. Extracellular ATP acts on Jasmonate Signaling to reinforce plant Defense. Plant Physiology. 2017;**176**(1):511-523. Available from: https://pubmed.ncbi.nlm.nih.gov/29180381

[18] Hermans C, Hammond JP, White PJ, Verbruggen N. How do plants respond to nutrient shortage by biomass allocation? Trends in Plant Science. 2006;**11**(12):610-617. DOI: 10.1016/j.tplants.2006.10.007

[19] Lambers H, Raven J, Shaver G, Smith S. Plant nutrient-acquisition strategies change with soil age. Trends in Ecology & Evolution. 2008;**23**(2):95-103. DOI: 10.1016/j.tree.2007.10.008

[20] Calvin M, Benson AA. The path of carbon in photosynthesis. Science (1979). 1948;**107**(2784):476-480. DOI: 10.1126/science.107.2784.476

[21] Zhu XG, Long SP, Ort DR. What is the maximum efficiency with which photosynthesis can convert solar energy into biomass? Current Opinion in Biotechnology. 2008;**19**(2):153-159. DOI: 10.1016/j.copbio.2008.02.004

[22] Schreier TB, Hibberd JM. Variations in the Calvin-Benson cycle: Selection pressures and optimization? Journal of Experimental Botany. 2019;**70**(6):1697-1701. Available from: https://pubmed.ncbi.nlm.nih.gov/30916343

[23] Bassham JA, Krause GH. Free energy changes and metabolic regulation in steady-state photosynthetic carbon reduction. Biochimica et Biophysica Acta (BBA) – Bioenergetics. 1969;**189**(2):207-221. DOI: 10.1016/0005-2728(69)90048-6

[24] Whitmarsh J, Govindjee. Photosynthesis. In: Digital Encyclopedia of Applied Physics. 2003. DOI: 10.1002/3527600434.eap327

[25] Spreitzer RJ, Salvucci ME. RUBISCO: Structure, Regulatory interactions, and possibilities for a better enzyme. Annual Review in Plant Biology. 2002;**53**(1):449-475. DOI: 10.1146/annurev.arplant.53.100301.135233

[26] Johnson X, Alric J. Central carbon metabolism and electron transport in *Chlamydomonas reinhardtii*: Metabolic constraints for carbon partitioning between oil and starch. Eukaryotic Cell. 2013;**12**(6):776-793. Available from: https://pubmed.ncbi.nlm.nih.gov/23543671

[27] Meloni M, Gurrieri L, Fermani S, Velie L, Sparla F, Crozet P, et al. Ribulose-1,5-bisphosphate regeneration in the Calvin-Benson-Bassham cycle: Focus on the last three enzymatic steps that allow the formation of Rubisco substrate. Frontier in Plant Science. 2023;**14**:1130430. Available from: https://pubmed.ncbi.nlm.nih.gov/3687559

[28] Rosa-Téllez S, Anoman AD, Flores-Tornero M, Toujani W, Alseek S, Fernie AR, et al. Phosphoglycerate kinases are Co-regulated to adjust metabolism and to optimize growth. Plant Physiology. 2017;**176**(2):1182-1198. Available from: https://pubmed.ncbi.nlm.nih.gov/28951489

[29] Zaffagnini M, Bedhomme M, Groni H, Marchand CH, Puppo C, Gontero B, et al. Glutathionylation in the photosynthetic model organism Chlamydomonas reinhardtii: A proteomic survey. Molecular Cell Proteomics. 2011;**11**(2):M111.014142-M111.014142. Available from: https://pubmed.ncbi.nlm.nih.gov/22122882

[30] Yamaoka C, Suzuki Y, Makino A. Differential expression of genes of the Calvin–Benson cycle and its related genes during leaf development in rice. Plant Cell Physiology. 2015;**57**(1):115-124. DOI: 10.1093/pcp/pcv183

[31] Sharkey TD, Bernacchi CJ, Farquhar GD, Singsaas EL. Fitting photosynthetic carbon dioxide response curves for $C_3$ leaves. Plant, Cell & Environment. 2007;**30**(9):1035-1040. DOI: 10.1111/j.1365-3040.2007.01710.x

[32] Cen YP, Sage RF. The regulation of Rubisco activity in response to variation in temperature and atmospheric CO2 partial pressure in sweet potato. Plant Physiology. 2005;**139**(2):979-990. Available from: https://pubmed.ncbi.nlm.nih.gov/16183840

[33] Pinna S, Kunz C, Halpern A, Harrison SA, Jordan SF, Ward J, et al. A prebiotic basis for ATP as the universal energy currency. PLoS Biology. 2022;**20**(10):e3001437-e3001437. Available from: https://pubmed.ncbi.nlm.nih.gov/36194581

[34] Biology, The Cell, Photosynthesis, using Light Energy to Make Organic Molecules. VIVA Open. 2023. Available from: https://vivaopen.oercommons.org/courseware/lesson/474/overview

[35] Blankenship RE. Molecular Mechanisms of Photosynthesis. Wiley; 2002. DOI: 10.1002/9780470758472

[36] Nelson N, Ben-Shem A. The complex architecture of oxygenic photosynthesis. Natural Review in Molecular Cell Biology. 2004;**5**(12):971-982. DOI: 10.1038/nrm1525

[37] Bonora M, Patergnani S, Rimessi A, De Marchi E, Suski JM, Bononi A, et al. ATP synthesis and storage. Purinergic Signaling. 2012;**8**(3):343-357. Available from: https://pubmed.ncbi.nlm.nih.gov/22528680

[38] Neupane P, Bhuju S, Thapa N, Bhattarai HK. ATP synthase: Structure, function and inhibition. Biomolecular Concepts. 2019;**10**(1):1-10. DOI: 10.1515/bmc-2019-0001

[39] Kaim G, Dimroth P. ATP synthesis by F-type ATP synthase is obligatorily dependent on the transmembrane voltage. EMBO Journal. 1999;**18**(15):4118-4127 Available from: https://pubmed.ncbi.nlm.nih.gov/10428951

[40] Daum B, Nicastro D, Austin J 2nd, McIntosh JR, Kühlbrandt W. Arrangement of photosystem II and ATP synthase in chloroplast membranes of spinach and pea. Plant Cell. 2010;**22**(4):1299-1312. Available from: https://pubmed.ncbi.nlm.nih.gov/20388855

[41] Kühlbrandt W. Structure and function of mitochondrial membrane protein complexes. BMC Biology. 2015;**13**:89. Available from: https://pubmed.ncbi.nlm.nih.gov/26515107

[42] Igamberdiev AU, Kleczkowski LA. Optimization of ATP synthase function

in mitochondria and chloroplasts via the adenylate kinase equilibrium. Frontier in Plant Science. 2015;**6**:10. Available from: https://pubmed.ncbi.nlm.nih.gov/25674099

[43] Feng TY, Yang ZK, Zheng JW, Xie Y, Li DW, Murugan SB, et al. Examination of metabolic responses to phosphorus limitation via proteomic analyses in the marine diatom Phaeodactylum tricornutum. Scientific Reports. 2015;**5**:10373. Available from: https://pubmed.ncbi.nlm.nih.gov/26020491

[44] Scheibe R. Maintaining homeostasis by controlled alternatives for energy distribution in plant cells under changing conditions of supply and demand. Photosynthetic Research. 2018;**139**(1-3):81-91. Available from: https://pubmed.ncbi.nlm.nih.gov/30203365

[45] Nikkanen L, Toivola J, Trotta A, Diaz MG, Tikkanen M, Aro EM, et al. Regulation of cyclic electron flow by chloroplast NADPH-dependent thioredoxin system. Plant Direction. 2018;**2**:11, e00093-e00093. Available from: https://pubmed.ncbi.nlm.nih.gov/31245694

[46] Mulo P. Chloroplast-targeted ferredoxin-NADP+ oxidoreductase (FNR): Structure, function and location. Biochimica et Biophysica Acta (BBA) – Bioenergetics. 2011;**1807**(8):927-934. DOI: 10.1016/j.bbabio.2010.10.001

[47] Pollak N, Dölle C, Ziegler M. The power to reduce: pyridine nucleotides--small molecules with a multitude of functions. Biochemical Journal. 2007;**402**(2):205-218. Available from: https://pubmed.ncbi.nlm.nih.gov/17295611

[48] Saint-Sorny M, Brzezowski P, Arrivault S, Alric J, Johnson X. Interactions between carbon metabolism and photosynthetic electron transport in a *Chlamydomonas reinhardtii* mutant without CO(2) fixation by RuBisCO. Frontier in Plant Science. 2022;**13**:876439. Available from: https://pubmed.ncbi.nlm.nih.gov/35574084

[49] Jacob J, Lawlor DW. Dependence of photosynthesis of sunflower and maize leaves on phosphate supply, ribulose-1,5-bisphosphate carboxylase/oxygenase activity, and ribulose-1,5-bisphosphate pool size. Plant Physiology. 1992;**98**(3):801-807. Available from: https://pubmed.ncbi.nlm.nih.gov/16668751

[50] Mettler T, Mühlhaus T, Hemme D, Schöttler MA, Rupprecht J, Idoine A, et al. Systems analysis of the response of photosynthesis, metabolism, and growth to an increase in irradiance in the photosynthetic model organism Chlamydomonas *reinhardtii*. Plant Cell. 2014;**26**(6):2310-2350. Available from: https://pubmed.ncbi.nlm.nih.gov/24894045

[51] Yamori W, Hikosaka K, Way DA. Temperature response of photosynthesis in C3, C4, and CAM plants: Temperature acclimation and temperature adaptation. Photosynthetic Research. 2013;**119**(1-2):101-117. DOI: 10.1007/s11120-013-9874-6

[52] Andersson I, Backlund A. Structure and function of Rubisco. Plant Physiology and Biochemistry. 2008;**46**(3):275-291. DOI: 10.1016/j.plaphy.2008.01.001

[53] Lorimer GH. The carboxylation and oxygenation of ribulose 1,5-bisphosphate: The primary events in photosynthesis and photorespiration. Annual Review in Plant Physiology. 1981;**32**(1):349-382. DOI: 10.1146/annurev.pp.32.060181.002025

[54] Raines CA. Increasing photosynthetic carbon assimilation in C3 plants to

improve crop yield: Current and future strategies. Plant Physiology. 2010;**155**(1):36-42. Available from: https://pubmed.ncbi.nlm.nih.gov/21071599

[55] Bassham JA, Calvin M. The way of CO2 in plant photosynthesis. Comparative Biochemistry and Physiology. 1962;**4**:187-204. DOI: 10.1016/0010-406X(62)90004-X

[56] Portis AR, Parry MAJ. Discoveries in Rubisco (Ribulose 1,5-bisphosphate carboxylase/oxygenase): A historical perspective. Photosynthetic Research. 2007;**94**(1):121-143. DOI: 10.1007/s11120-007-9225-6

[57] Erb TJ, Zarzycki J. A short history of RubisCO: The rise and fall (?) of Nature's predominant CO(2) fixing enzyme. Current Opinion in Biotechnology. 2017;**49**:100-107. Available from: https://pubmed.ncbi.nlm.nih.gov/28843191

[58] Portis AR Jr. The rubisco Activase-rubisco system: An ATPase-dependent association that regulates photosynthesis. In: Annual Plant Reviews Book Series, Volume 7: Protein-Protein Interactions in Plant Biology. John Wiley & Sons, Ltd; 2018. pp. 245-259. DOI: 10.1002/9781119312994.apr0057

[59] Hammond ET, Andrews TJ, Woodrow IE. Regulation of ribulose-1,5-bisphosphate Carboxylase/Oxygenase by carbamylation and 2-carboxyarabinitol 1-phosphate in tobacco: Insights from studies of antisense plants containing reduced amounts of rubisco activase. Plant Physiology. 1998;**118**(4):1463-1471. Available from: https://pubmed.ncbi.nlm.nih.gov/9847122

[60] Scheibe R. Malate valves to balance cellular energy supply. Physiology Plant. 2004;**120**(1):21-26. DOI: 10.1111/j.0031-9317.2004.0222.x

[61] Wedel N, Soll J. Evolutionary conserved light regulation of Calvin cycle activity by NADPH-mediated reversible phosphoribulokinase/CP12/ glyceraldehyde-3-phosphate dehydrogenase complex dissociation. Proceedings of thee National Academy Science USA. 1998;**95**(16):9699-9704. Available from: https://pubmed.ncbi.nlm.nih.gov/9689144

[62] Foyer CH, Lopez-Delgado H, Dat JF, Scott IM. Hydrogen peroxide- and glutathione-associated mechanisms of acclimatory stress tolerance and signalling. Physiology Plant. 1997;**100**(2):241-254. DOI: 10.1034/j.1399-3054.1997.1000205.x

[63] Meng X, Chen WW, Wang YY, Huang ZR, Ye X, Chen LS, et al. Effects of phosphorus deficiency on the absorption of mineral nutrients, photosynthetic system performance and antioxidant metabolism in *Citrus grandis*. PLoS One. 2021;**16**(2):e0246944-e0246944. Available from: https://pubmed.ncbi.nlm.nih.gov/33596244

[64] Mihretie FA, Tsunekawa A, Haregeweyn N, Adgo E, Tsubo M, Masunaga T, et al. Exploring teff yield variability related with farm management and soil property in contrasting agro-ecologies in Ethiopia. Agricultural System. 2022;**196**:103338. DOI: 10.1016/j.agsy.2021.103338

[65] Strand A, Hurry V, Henkes S, Huner N, Gustafsson P, Gardeström P, et al. Acclimation of Arabidopsis leaves developing at low temperatures. Increasing cytoplasmic volume accompanies increased activities of enzymes in the Calvin cycle and in the sucrose-biosynthesis pathway. Plant Physiology. 1999;**119**(4):1398-1387. Available from: https://pubmed.ncbi.nlm.nih.gov/10198098

[66] Lawlor DW. Limitation to photosynthesis in water-stressed leaves: Stomata vs. metabolism and the role of ATP. Annals of Botany. 2002;**89**(7):885-871. Available from: https://pubmed.ncbi.nlm.nih.gov/12102513

[67] de Bang TC, Husted S, Laursen KH, Persson DP, Schjoerring JK. The molecular–physiological functions of mineral macronutrients and their consequences for deficiency symptoms in plants. New Phytologist. 2020;**229**(5):2446-2469. DOI: 10.1111/nph.17074

[68] von Wandruszka R. Phosphorus retention in calcareous soils and the effect of organic matter on its mobility. Geochemical Transaction. 2006;7:6. Available from: https://pubmed.ncbi.nlm.nih.gov/16768791

[69] Adjei-Nsiah S, Alabi BU, Ahiakpa JK, Kanampiu F. Response of grain legumes to phosphorus application in the Guinea Savanna agro-ecological zones of Ghana. Agron Journal. 2018;**110**(3):1089-1096 Available from: https://pubmed.ncbi.nlm.nih.gov/33281193

[70] Mehra P, Pandey BK, Giri J. Improvement in phosphate acquisition and utilization by a secretory purple acid phosphatase (OsPAP21b) in rice. Plant Biotechnological Journal. 2017;**15**(8):1054-1067. Available from: https://pubmed.ncbi.nlm.nih.gov/28116829

[71] Srivastava S, Ranjan M, Bano N, Asif MH, Srivastava S. Comparative transcriptome analysis reveals the phosphate starvation alleviation mechanism of phosphate accumulating *Pseudomonas putida* in *Arabidopsis thaliana*. Scientific Report. 2023;**13**(1):4918. Available from: https://pubmed.ncbi.nlm.nih.gov/36966146

[72] Finkelstein R. Abscisic acid synthesis and response. Arabidopsis Book. 2013;**1**(11):e0166-e0166. Available from: https://pubmed.ncbi.nlm.nih.gov/24273463

Chapter 4

# Sustainable Management of Phosphorus in Agriculture for Environmental Conservation

*Tahsina Sharmin Hoque, Deepranjan Sarkar, Rahul Datta, Mohammad Golam Kibria, Rafi Ullah, Nazeer Ahmed, Mohammad Anwar Hossain, Asim Masood and Naser A. Anjum*

## Abstract

Phosphorus (P) is an essential macronutrient for plant growth and development. Although the P-concentration in soil is 1000 folds higher than in plants, it is rarely available for plant uptake due to low diffusion and high fixation rate in soil. Hence, plants experience P-deficiency in the absence of P-fertilization, which may cause approximately a 30–40% decrease in crop yield. This highlights the importance of using a large amount of phosphate fertilizers to meet crop demands. As P-fertilizer is derived from a nonrenewable and finite source of rock phosphate, this resource is decreasing over time. In addition, farmers are applying P-fertilizers randomly without considering the soil stock, which leads to the loss of P-resources. The low P-use-efficiency (PUE) of plants in the field condition (15–20%) highlights that most of the soil-applied P remains unavailable to plants, and excess P causes ground and surface water contamination (i.e., eutrophication) through leaching and runoff, which ultimately results in environmental pollution. Therefore, it is crucial to apply P-fertilizers considering the soil test value and PUE to protect the environment from contamination and sustainable management of P-resources. This chapter mainly focuses on the sustainable management of P in agricultural fields for environmental conservation.

**Keywords:** plant phosphorus nutrition, phosphorus use efficiency, phosphorus fertilizers, environmental pollution, sustainable management

## 1. Introduction

Phosphorus (P) is an essential macronutrient for plant growth and development and is also considered as an important nonrenewable global resource for the agricultural production system [1]. It is an important growth-limiting factor in the agricultural production of the world due to its immobility and poor availability to plants in soil [2, 3]. Continuous application of P-fertilizer is crucial for the modern agricultural production system. A large amount of P-fertilizer is derived from the finite and nonrenewable source of phosphate rock. If this situation continues, it is predicted that the current global rock phosphate reserves could run out within the next 50–100 years [4, 5], or peak phosphorus could occur due to the global demand for agricultural production within a decade [5]. The fertilizer industry is concerned with the already rising cost of phosphate fertilizer associated with agricultural production. Because of this, scientific researchers are paying more attention to the long-term sustainable management of P-fertilizer. It is evident from different research findings that P is involved in several plant physiological and biochemical functions of plants. It is actively involved in growth and reproduction [6], flowering [7], seed formation [8], root development [9], disease resistance [10] and photosynthesis [11]. As a structural component of nucleic acids, lipids, and sugars within the plant cells, it is actively involved in the growth and development of plants both at the cellular and whole plant levels. The deficiency of P in the soil causes a significant decrease in primary root growth. It alters the root architecture (morphology, topology, and root distribution) by lateral root formation and increasing the length and density of root hairs [12]. Reduced root growth also causes a decrease in plant growth and results in significant crop loss. This highlights the importance of P-nutrition in crop production and emphasizes sustainable P-fertilization to enhance plant growth and development.

Phosphorus is a scarce and complicated element because it involves organic and inorganic (35–70% of total P in soil) stocks in soil [13]. P is one of the most important nonrenewable resources, and it has already received global attention due to its low availability in soil (mainly due to slow diffusion and high fixation in soil) [14]. Optimum P-fertilization is necessary to make an economic profit in farming, but the application of P-fertilizer to soils having excess P is not economically beneficial. If P is not applied to soil, plants take up P from soil reserves. In contrast, P applied to the low P soil might be profitable in farming [15]. Soils containing optimal amounts of P need to be properly managed, with appropriate fertilizer applications to maintain (or slightly decrease) their P-status. Adopting sustainable P-fertilizer management to reduce fertilizer costs would also decrease the negative effects of leaching and runoff to the environment.

Therefore, this chapter will focus on different aspects of P-management for sustainable environmental conservation. It will also review the biochemical functions, forms, behavior, and transport of P in soil. The problems associated with P-management strategies and possible action plans for sustainable agricultural and environmental conservation regarding the changing global perspectives are presented.

## 2. Biochemical and physiological functions of phosphorus

Phosphorus is a major essential nutrient element for plants that plays a crucial role in every developmental stage of life from very early to adult. This element improves crop yield and quality, and the functions of P cannot be performed by any other nutrient [14]. A plant can neither complete its normal reproductive process nor reach its highest yield potential without an adequate supply of P because this macronutrient has a major contribution to multiple cellular functions. As a structural component of nucleic acids, sugars, and lipids, P takes part in all the growth and developmental processes at both cellular and whole plant levels as, for example, cell division/elongation, membrane structure maintenance, enzyme activation, or inactivation, biomolecule synthesis, photosynthesis, respiration, high-energy molecules formation such carbohydrate metabolism, nitrogen fixation, seed germination, seedling establishment, root and shoot development, flower and seed formation, improves disease and stress resistance etc. (**Figure 1**).

Various biochemical functions of P are discussed in the following sections.

## 2.1 Structural component

The total P-concentration determines P compartmentalization within the plant cells. P makes up about 0.2% of a plant's dry weight [16, 17]. The optimum concentration for most of the crops is below 4.0 mg P $g^{-1}$ dry weight [18–20]. Two forms of P function in plant tissues such as free inorganic orthophosphate form (Pi) and organic phosphate esters. Pi is metabolically active and is located in the cytoplasm. When P is excess, it is stored in the vacuole in esterified forms, and from this organelle P is supplied to cytoplasm on cellular demand. Nucleic acids, phospholipids, phosphorylated metabolites, and proteins are some of the esterified forms of P.

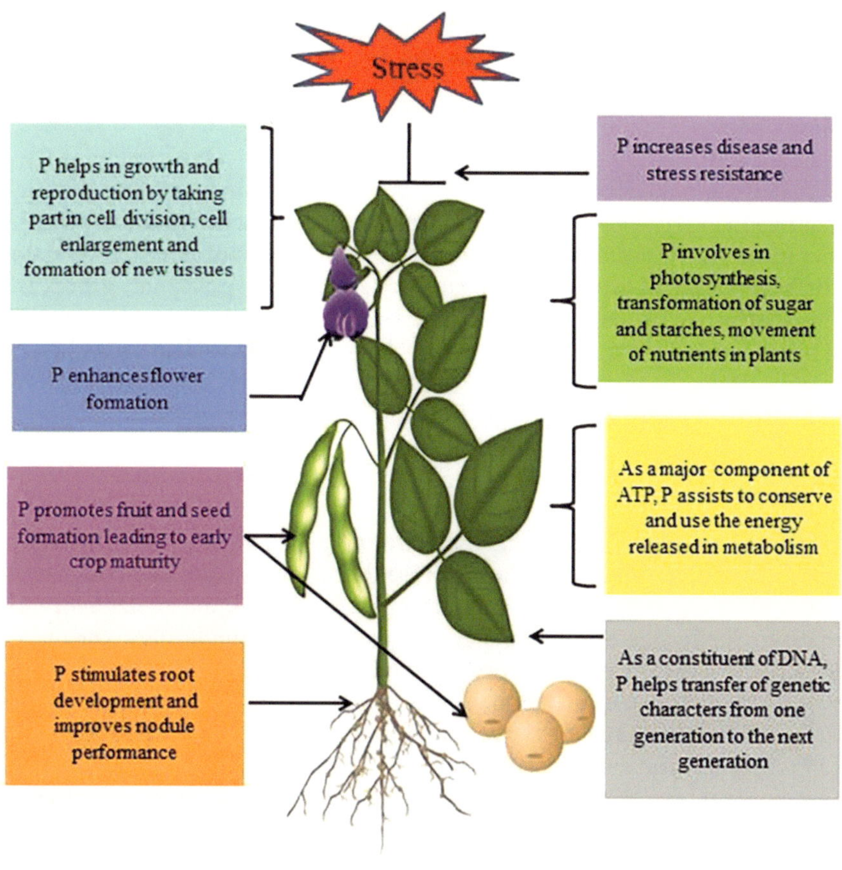


**Figure 1.**
*Schematic representation of the major physiological and biochemical functions of phosphorus in a typical legume plant. P has a significant contribution to photosynthesis, carbohydrate metabolism, energy release, transfer of nutrients and genetic materials, resistance to stresses and diseases, cell division and development, formation of plant organs such as flower, fruit, and seed, development of roots, nodules, etc., and thus being an indispensable element in plants life from germination up to maturity.*

### *2.1.1 Sugar phosphates*

Sugar phosphates, most of which are Pi esters of monosaccharides, are intermediate compounds of carbohydrate metabolism. Phosphorylation of carbohydrates followed by reaction with ATP forms sugar phosphates, some of which include phytic acid, glucose-6-phosphate, and dihydroxyacetone phosphate. As constituents of glycolysis and respiratory reactions, these phosphorylated compounds play significant roles in photosynthesis as well as the synthesis and breakdown of carbohydrates.

### *2.1.2 Phospholipids*

Phospholipids play a pivotal role in the structural regulation and dynamics of the cytoskeletal organization by the interaction of its molecules with multiple actin-binding proteins, viz. profilin, gelsolin, α-actin, cofilin, filamin, vinculin, etc. [21, 22]. Phospholipid molecules are considered as essential structural elements of cell membranes, emerging as important second messengers to regulate plant growth, development, and cellular responses to environmental stresses through different mechanisms [23]. It is well recognized that phosphoinositides are involved in membrane trafficking and signaling in the cell responses to stress stimuli such as salt, osmotic, temperature, pathogen stressors, etc. [23, 24].

### *2.1.3 Nucleic acids*

As a major component of nucleic acids, P is involved in the transfer of genetic materials needed for growth and development from one generation to the succeeding generation. Nucleic acids are polynucleotides that consist of nitrogen-containing aromatic bases attached to a pentose sugar, connected with a phosphate group. DNA and RNA are the two principal classes of nucleic acids that form the largest organic P pool in plants, ranging from 0.3 to 2.0 mg P $g^{-1}$ dry weight in various crops [25]. Both nucleic acids direct the process of new protein synthesis such as enzymes and determine the inherited traits in living organisms including plants.

## 2.2 Energy and electron transport

The role of P in cellular metabolism, mainly energy transport, is of great significance. P is involved in various cellular processes because it is linked with high-energy bonds, and the high-energy P-containing compounds such as phosphoanhydride, acyl phosphate, enol phosphate, etc., transfer energy to acceptor molecules. The phosphates are formed at the sites of energy production and are donated to other molecules at sites of energy consumption. The most common P energy currency is found in ATP, whose hydrolysis releases a great amount of free energy required for multiple cellular processes like macromolecule synthesis, membrane phospholipid formation, nutrient transport, etc. For example, α, β, and γ-phosphate release energy of 13.8, 27.2, and 34.0 kJ $mol^{-1}$, respectively, after hydrolysis [25]. Similarly, phosphoanhydride bonds are found in di- and triphosphate molecules in guanine, cytosine, uracil, and thymine nucleosides. In gluconeogenesis and saccharide metabolism, guanosine triphosphate (GTP) and uridine triphosphate (UTP) are important electron donors. Phosphorus is an essential part of the structure of triphosphopyridine nucleotide (TPN) that provides a similar transportation function in plants as adenosine diphosphate (ADP) and adenosine triphosphate (ATP). These nucleotides function as carriers of electrons or hydrogen between sites of oxidation and reduction reactions occurring in photosynthesis and respiration.

### 2.3 Photosynthesis

Phosphorus directly or indirectly regulates the photosynthesis process as a primary substrate, utilizing light energy to form sugars and a three-phosphate-containing molecule ATP in the presence of chlorophyll, $CO_2$, and water. It should be noted that ATP is the key driver of different metabolic reactions in plant cells that is required for the development of structural and storage components. P is a prime element of the thylakoid membrane and is crucial for the regulation of photosynthetic machinery, viz. PSI, PSII, LHCP, cyt-f, cyt-b, and antenna mobility, as reported by Rychter and Rao [26]. In fact, apoprotein phosphorylation of antenna is a vital step in photosynthesis. Various processes involved in photosynthesis depend largely on the availability of Pi, which is controlled by an increase in photorespiration. Alterations in Pi availability in cytoplasm modulate the activation of multiple enzymes (such as RuBisCO, sedoheptulose-1,7-bisphosphatase, and fructose-1,6-bisphosphatase) and amounts of intermediates of the photosynthetic carbon reduction cycle. Besides, Pi possesses a great role in the partitioning of the photosynthates in various plant tissues and the distribution of newly fixed C between chloroplasts and cytoplasm during starch and sucrose synthesis, respectively.

### 2.4 Glycolysis

Phosphorus is involved in every reaction of the glycolysis process in plants. In the initial reactions of glycolysis, energy from ATP is necessary. However, in the subsequent reactions, ATP is generated with a net release of energy. Actually, ATP is the energy end product of glycolysis, which is the predominant pathway of carbohydrate metabolism in plant tissue. Pentose phosphate pathway is one of the major aerobic processes of carbohydrate degradation and synthesis where a series of oxidation-reduction reactions begin with five carbon sugars having a phosphate molecule attached to them. The activities of several glycolytic enzymes phosphofructokinase (PFK), nicotinamide adenine dinucleotide (NAD)-glyceraldehyde 3phosphate dehydrogenase (G3PDH), 3-Phosphoglyceric acid (3-PGA) kinase, and pyruvate kinase (PK) depend on the concentration of adenylate and Pi. The levels of respiratory intermediates, viz. hexose phosphates and 3-phosphoglyceric acid (PGA), reduce during P-deficiency. The activities of PFP and nonphosphorylating NAD-G3P-DH, phosphoenolpyruvate (PEP) carboxylase, and PEP phosphatase have been found to increase under P-deprived conditions.

### 2.5 Nutrient transport

Plant roots extract nutrients from soil solution and translocate them within the body, and most of this movement depends on a transport system through the cell membrane that requires energy to override the opposing force of osmotic equilibrium, which is provided by ATP through its high-energy phosphate bonds. ATP or some other phosphorylated compounds conduct the transport of nutrients through membranes of plant organs such as roots and leaves by an energy-driven process.

Several plasma membrane-bound transport proteins play vital roles in nutrient transport, and for example, proton pump named $H^+$-ATPase imports nutrients into the plant together with the export of $H^+$ by utilizing ATP [27]. Therefore, to enable nutrient transport smoothly with the use of ATP, these membrane proteins are abundant in the epidermal and endodermal root cells, xylem and phloem cells [28–30].

### 2.6 Nitrogen fixation

Phosphorus assists in enhancing rhizobial bacterial population in soil. It is considered as a principal plant nutrient that is strongly involved in biological nitrogen fixation and nodule performance in legumes. *Rhizobium* bacteria require energy for growth, reproduction, and functioning, which is provided by ATP that is transformed into ADP with concurrent liberation of inorganic phosphate. For each molecule of $N_2$ reduction for fixation, at least 16 molecules of ATP are hydrolyzed, showing that N-fixation is an energy-demanding process. Nodule formation and development are largely inhibited in P deficit condition, and as a result, nitrogen fixation is greatly impaired. Nitrogenase enzyme activity is also positively correlated with plants P nutritional status. Researchers found higher root nodule biomass in pea plants with higher supply of P [31] and lower concentration of ATP and energy charge in soybean nodules with a low P supply [32].

### 2.7 Seed germination

The embryo in plant seeds requires P for growth and development until the formation of a root system. For seed germination and seedling vigor, the seed P-content is a vital factor because seed P is the only P available to plants during germination that aids in nutrition and faster establishment of young seedlings. With the rise of P-concentration in seed, dry matter production of seedlings of annual plants increases. P quantity in seeds of plants differs from about 0.15 to 0.60% in seeds of wheat, oats, barley, and lupins and from 0.30 to 1.10% in those of subterranean clover and annual medics [33]. Conversely, according to some recent studies [34, 35], a lower concentration of seed P exerted no difference in seedling vigor, plant biomass, and yield in comparison with a high concentration of seed P, suggesting that an optimum concentration of seed P is adequate for seed germination whereas, higher P-concentration in seeds might be useless.

### 2.8 Root and shoot growth

Phosphorus is inextricably involved in plant growth from cellular to whole plant level, playing a significant role in cell division and cell enlargement that are crucial for root and shoot development. It stimulates root development required for the plant to get nutrients and support from the soil. Phosphorus is essential for better growth parameters of plants such as height, biomass of shoot, area and number of leaves, and timely appearance and development of tillers in cereals [36]. The deficiency of P can lead to limited cell divisions and enlargement in leaves, resulting in reduced shoot biomass. Plant growth is considered to be more sensitive to P-availability compared to photosynthesis [37]. Under long-term Pi-deficient conditions, the relative growth rate decreases as a result of reduced ATP concentration in roots [38]. Limited leaf expansion under low P-supply might be due to low turgor pressure for leaf expansion resulting from decreased water transport in the leaf from the growth medium.

### 2.9 Flower, fruit, and seed formation

Phosphorus is responsible for the reproductive development of plants such as formation of flower, fruits, and seeds as well as crop maturity at the right time. P-deficient plants take longer period to mature and bear few fruits and seeds with poor quality. Phosphorus nutrition can regulate anthocyanin production in leaves flower stalks by modulating the activities of phenylalanine ammonia-lyase (PAL) and chalconeisomerase (CHI) and epidermal pH values [39]. Optimum amount of P in soil ensures a higher number, dry matter, and yield of fruits and seeds with a greater harvest index. For development and maturation, both fruits and seeds require large quantities of P, and an inadequate supply of P can reduce the size, number, viability, and quality of seeds. In cereal crops such as rice, wheat, and maize, the majority of total P (about 75%) is stored as phytin or associated compounds, whereas 4–9% is stored as inorganic phosphate and 15–25% as cellular-P [40].

### 2.10 Disease and stress resistance

Phosphorus is crucial for the general health and vigor of plants, and it enhances resistance and tolerance to stress and diseases, which otherwise can reduce crop quality and production. P-balancing with other nutrient inputs is of prime importance to decline the risk of disease infection. In a complete and balanced fertility program, P is a major element that can improve crop health to withstand stress and to become less susceptible to disease occurrence. However, the role of P in disease resistance is variable and inconsistent [41].

## 3. Forms of phosphorus in soil

The total P in soil is a combination of both inorganic and organic forms. Generally, the inorganic P is higher than the organic P because it accounts for around 50–85% of soil total P [42]. The inorganic P constitutes some active forms of P like calcium P (Ca-P), iron P (Fe-P), and aluminum P (Al-P) and inactive forms like occluded P, reductant soluble P, and residual P [43, 44]. The concentration of total P ranges from 50 to 3000 ppm in soils, but it has little contribution to the bioavailability of P in soils [45]. The solution P, which meets the plant's requirement, is available in very low concentrations (around 0.05 ppm) in soil [46].

Organic P consists of inositol phosphates, phospholipids, and nucleic acids. Inositol phosphates exist as a dominant pool of organic P in soil because the other two forms are highly susceptible to microbial degradation [47]. Globally, the share of organic P in total soil P is observed between 1 and 30% [48, 49]. In Indian soils, this share varied from 10 to 50% [50]. However, the contribution of organic P may increase up to 80% in forest and grassland ecosystems [51].

## 4. Behavior and chemistry of phosphorus in soil

When P-fertilizers are applied to the soil, they react with cations (Ca, Fe, and Al) and form insoluble P-compounds. Consequently, P-use-efficiency declines, and crops can recover only 15–20% of applied P [52]. In mineral soils, P is highly immobile [53]. P-cycling in soil occurs through numerous processes, viz., mineralization-immobilization, precipitation, adsorption-desorption, dissolution, and plant uptake [46, 54]. The fixation of P is common in acid soils because P precipitates as oxides and hydrous oxides of Fe and Al or gets adsorbed on clay surface or surface of oxides and hydrous oxides of Fe and Al [55, 56]. Similarly, in calcareous soils, these phenomena (precipitation and adsorption) lead to the formation of insoluble Ca compounds [57, 58]. The adsorption process takes place by ligand exchange, which favors monodentate phosphate complexation [54, 59]. Based on adsorption, P-forms can be classified as labile and nonlabile [60]. Labile forms are related to weak adsorption because of low desorption time. They supply P to the soil solution. Thus, under equilibrium conditions, P-dynamics in the soil [54] are represented as:

$$\text{Nonlabile P} \leftrightarrow \text{Labile P} \leftrightarrow \text{solution P}. \quad (1)$$

In the mineralization process, the soil organic P is converted into inorganic P by the microbes present in the soil. This is also known as P solubilization. Organic P hydrolysis is performed by microbes producing enzymes (phosphatase, phytase) and organic acids (formic, oxalic) [61]. When the available P is low in the soil, microbial immobilization of P begins, i.e., inorganic P gets transformed into organic P [62].

With the increase in soil age, the concentration of nonlabile forms of P increases [44]. P-availability in soil is dependent upon pH, organic matter, clay content and/or type, moisture content, temperature, aeration, and other properties of the soil. In heavily P-fertilized soils of northern Iran, the reported higher P-adsorption was argued due to the high content of clay and Fe and Al oxides [43]. The capacity of soils to supply this nutrient to plants or the buffering capacity of soils is determined by quantity and intensity factors [63]. This is determined by sorption and desorption curves. Quantity is the solid phase that equilibrates with the solution and can also represent the buffering capacity of soils to meet the P requirement of plants [64]. Therefore, to improve the efficacy of P-fertilizer recommendations, buffering indices must be considered in our soil testing programs to know about P-uptake at different stages of plant growth [65]. The release of P occurs in simultaneous processes, viz., desorption at a higher rate and diffusion at a slower rate [66].

The availability of P increases with the addition of organic matter. As through mineralization of organic matter, the available form of phosphorus is released to soils. Retention of P reduces because phosphate absorbed into soil competes with organic molecules. In soils with high clay contents, the retention capacity of phosphorus will be high because the surface area per unit volume is very high of clay particles that absorb phosphorus very easily. On the other hand, the adsorption capacity of P increases when soils have a sufficient amount of minerals. The mineral composition of the soil influences the phosphorus adsorption capacity. The availability of P will be higher in soils that have 6–7 pH. At low soil pH, the Fe and Al make strong bonds with phosphorus, and phosphate tends to precipitate with calcium at high soil pH. Another factor also affects the availability of phosphorus to plants; like in cool weather, the organic matter takes a long time to decompose compared with warm and hot weather.

## 5. Phosphorus transport and contamination in the environment

Agricultural land is the major source of P-transport in surface waters. The loss of P as runoff causes P enrichment or eutrophication in aquatic ecosystems.

Therefore, agricultural management practices should focus on reducing P-transport caused by excessive consumption of inorganic fertilizers and organic manures [67]. Both soil and hydrological factors are responsible for the movements of P [68]. The former decides the initial transport, while the latter is associated with transport as well as the pathways. There are three major pathways by which P is transported to surface water bodies. Those are surface runoff, subsurface flow, and vertical flow [67]. Minimizing soluble P concentrations or increasing water infiltration in soil for P-adsorption will help in decreasing the dissolved P-transport [69]. The subsurface flow of P is a concern only for soils having high P-saturation or preferential flow. At high rainfall intensity (hydrological factor), fertilizers containing a higher concentration of water-soluble P can increase P-transport in runoff [70]. Manures containing high organic P (like poultry manure) may contribute to dissolved P-losses via leaching and displacement of loosely bound inorganic P [71, 72]. The possible transport of P in the environment has been highlighted in **Table 1**.

Long-term application of P-fertilizers (like triple superphosphate) may increase the concentrations of some trace elements or heavy metals (arsenic, cadmium, copper, chromium, nickel, vanadium, and zinc) in soil [73]. Mining and processing of rock phosphate cause significant contamination (from radioactive to heavy metal pollutants) in the environment [74, 75]. Phosphogypsum, a by-product obtained during the processing of rock phosphate, marks a serious potential hazard for human health and soils, water, and atmospheric pollution [75]. The use of organic wastes (pig slurry and cattle slurry) above crop P demands saturates soil adsorption sites, leading to P-migration in the soil profile and subsurface water contamination [76]. Therefore, instead of the bulk application of the bioresources, technologies should be framed to recover P from human and animal wastes including sewage, sludge, manure, incineration ashes, etc., by developing contaminant-free fertilizers, e.g., struvite (ammonium magnesium phosphate) and/or other soluble products for efficient P-cycling [77].

| P-sources | Anthropogenically contribution per day | Contribution percent of total P (%) |
|---|---|---|
| Kitchen waste disposal | 0.1 | 4 |
| Toilet | 1.6 | 59 |
| Bath, sinks, and machines | 1 | 37 |
| Total phosphorus | 2.7 | — |
| **Water body** | **Water body type** | **% Total P load from septic systems** |
| Cottonwood Lake | Freshwater lake | 4 |
| Lake Carlton | Freshwater lake | 14 |
| Pine View Reservoir | Reservoir | 20 |
| Pend Oreille Lake | Freshwater lake | 25 |
| Ontario Lakes | Freshwater lake | 55 |
| Otsego Lake | Freshwater lake | 20 |

**Table 1.**
*Transport of phosphorus in the environment.*

## 6. Importance of assessing phosphorus use efficiency

Using phosphorus efficiently means that the phosphorus taken up by plant per unit produces yield or biomass [78] and is distinguished when relevant by using subscripts, PUEt and PUEy, respectively. Phosphorus use efficiency may improve through strong agronomic practices, like the calculated amount of fertilizers, suitable timing, right place, and right crops. Mostly, the biomass for any plant is taken from the above-ground parts. There are so many limitations to reserves rock phosphate globally, and awareness is increasing. These reserves are used to increase or maintain the present agricultural productivity and produce crops [5]. Around 57% of the annual grain crops like pulses, cereals, and oil seeds cover the dietary energy for the present world's rapid growing population [79].

The primary goal of testing soils for P is to identify the supplementary P needed to prevent crop losses due to P-deficiency. Plant-available phosphorus in soil can be estimated using a soil test that offers an index of plant-available P. Second, soil testing for P is used to track the amount of accessible P in the soil. For evaluating fertilization procedures and the selection of waste disposal options, the data can be valuable.

A wide variety of chemical forms of P can be found in soil. All of these factors influence the plant-available pool to variable degrees. The amount of plant-available P in a given soil is not a fixed figure. According to a variety of soil and plant root properties and the environment, this can vary. A soil's plant-available P-content can be challenging to forecast. However, some effective P extraction methods have been established, which correlate well with P-uptake in controlled conditions. Routine soil fertility tests typically refer to a relatively rapid nutrient extraction, which results in an accessible soil nutrient value connected with crop response to fertilization. It is usual to practice to employ Mehlich-1 (M1) and Mehlich-3 (M3) for fertilizer P and K rate recommendations. There may be some variations in fertilizer rate recommendations even if the numerical soil test findings are the same, even if soil testing laboratories utilize equivalent extraction and quality control techniques and comparable instruments [80–82].

Soil testing relies heavily on developing standardized techniques (extract and analytical methodologies), test interpretation, and nutrient recommendations, all of which are based on field calibration and validation.

The following documents contain extensive information on soil testing, soil test extractant, and the correlation and calibration processes, as well as fertilizer recommendation regulations:

1. First, we need to understand what soil testing for plant-available nutrients means and why we use it in the first place.

2. Processes for the Correlation and Calibration of a Soil Test Extractant.

3. Recommendation Philosophies for the use of fertilizer.

4. Extracting Nutrients from Acid-Mineral Soils of Florida using Mehlich-3 Reagent.

Test results for specific soil features and P-availability are affected by some different soil parameters.

There is a perception that carbonates in soils pose a barrier to robust acid extraction procedures like Bray and Kurtz or Mehlich I. The Bray PI test for calcareous soils is lower than the $NaHCO_3$ test [83]. $CaCO_3$ neutralizes the acid, releasing Ca, which precipitates the fluoride. This reduction has largely been attributed to this process. Thus, the ability of the extractant to remove P is diminished. However, the Bray PI and $NaHCO_3$ tests have been performed equally on calcareous soils in Colorado and Nebraska in other correlation investigations [84]. As a result, factors other than the total $CaCO_3$ level may be involved in deciding whether the acid tests are unsuccessful. The use of acid tests on calcareous soils is generally avoided because of these additional unknown issues. Soil-to-solution ratios of 1:100 or 1:100 have enhanced correlations on neutral and calcareous soils and appear to alleviate the problem of soil toxicity [85–87].

The pH of noncalcareous soils has been identified as another factor influencing soil test performance, although not consistently. A field study in British Columbia showed that correlations were higher for alkaline soils than acid soils for both Bray PI and $NaHCO_3$ tests [88]. A South Dakota study showed that correlations were lowest in the pH range of 6.6–7.0 for both tests [86]. An Ohio study demonstrated only slight reductions in correlations when the pH exceeded 5.5 for Bray PI and $NaHCO_3$ tests [89].

## 7. Effect of soil properties on phosphorus use efficiency (or phosphorus dynamics)

The efficiency of P-use depends mostly on the complex interactions among the physical, chemical and biological characteristics of soils, and the processes involved with these properties. These properties can directly influence root growth and development, restricting P-uptake. Soil depth, structure, stoniness, moisture retention, and composition of the soil atmosphere can influence P-dynamics. Notably, the prevailing soil properties such as soil texture, mineralogical composition, total surface area, pH, $CaCO_3$ content, organic matter content, the presence of Fe and Al-hydroxides, etc., can considerably alter the solubility, availability, and extractability of P in soil [90]. Due to the presence of dense subsurface layers (e.g., plow pans) and surface-soil compaction, the diffusion of phosphate ions in soil is decreased, and root growth and development are hindered. Although the slow movement of P by diffusion is usually attributed to the tortuosity of the pore system, the reactive sites for P-adsorption on soil minerals around the pores can hold phosphate ions, temporarily or permanently, that slow or prevent their movement along the pore [91]. Soil texture influences the chemical behavior and hydrology of soils and affects the formation of Al-organic bonded stable P and leaching of P from soils [92, 93]. By practicing zero-tillage, cultivating permanent crops, and maintaining a crop-residue cover, and by minimizing traffic over the soil surface and reducing invasion of livestock in wet and heavy-textured soils in humid temperate regions, the physical limitations of P-uptake can be controlled to some extent. Besides, soil-borne fungal pathogens and nematodes injure plant roots and limit P-uptake and to overcome this problem, crop rotation is a suitable option.

Soil acidity is a chemical limitation to the efficient use of P-fertilizers as it has adverse effects on P-uptake. Soil pH regulates the release of Al from various clay minerals and the dissolution of Al hydroxy compounds in soil. At low pH with high Al-concentrations in the soil solution, root tips and lateral roots are thickened and turn brown, causing the reduction in P-uptake. As a result, P translocation in the upper parts of the plant decreases, and P metabolism is hampered. These problems can be ameliorated by adding lime and other acid-neutralizing materials that can increase pH as well as base saturation percentage. P-distribution in different soil P pools is an inherent soil property, and changes in the P-distribution in the pools are difficult to achieve. Scientists around the globe investigated the impact of pH change by adding lime to acid soils on retention and extractability of P and found no consistent influence on soil P availability [94–97]. The amount of readily available plant P increases with the increase of organic matter in the soil [98]. In P-deficient soils, applying a material that can compete with the phosphate ion for the adsorption sites within the soil such as silica or silicate might be a suitable option to increase P-availability [99].

## 8. Modeling phosphorus transport to water bodies and phosphorus index

According to the conceptual model and hillslope hydrology, it is vital to identify significant source locations, which are areas with disproportionately high P-losses, to efficiently manage P-movement regularly at the field and watershed scales. Soil and water assessment tools like the Soil and Water Assessment Tool [100] have been developed to identify critical source locations of P-migration (SWAT).

While various developments in modeling, including graphical user interfaces and geographic information system (GIS) layers, have been made in the field, the fundamental methodologies to modeling P-movement have remained essentially unchanged [101]. Land use, soil texture, and topography are the primary inputs for most P-transport models, but other factors like management techniques are also widely used as data. There are three preliminary modeling approaches: processed-based, export-coefficient, and statistical models [102].

For the original P-index, Lemunyon and Gilbert [103] set out to assess the risk of P-transport to water bodies, identify the essential components that drive P-loss, and assist in selecting management measures that reduce P-loss. It has evolved significantly since its introduction in 1993 [103] from being a critical source area identification to now serving as best management practice selectors in manure application scheduling tools, manure application rate calculators, and regulatory mechanisms of some states in the United States of America [104]. In the United States and Europe, P-indices have included as many as 34 variables from each location [105]. Runoff class, soil erosion, irrigation erosion, soil P-test, and P-fertilizer application rate and technique, organic P-source application rate, and method of organic P-source application were the initial eight parameters to evaluate in the original P-index, which was updated in 2010.

Some site factors may be more important than others in influencing P-migration from the site. The weighting factor for each site attribute was determined by professional judgment and historical experience. To get a weighted score for a site characteristic, the P-loss rating value is first multiplied by the site characteristic weighing factor. P-loss is less likely in fields with low P-indices, while P-loss is more likely in areas with high P-indices.

## 9. Global phosphorus consumption

Due to the growing world population and increasing demand for food consumption, the growing demand for bioenergy crops will affect the future P-use particularly

| Year | Diet | P-consumption (kg/capita/y) | | Total global P-consumption (Mt) | Fertilizer P-use (Mt) |
|---|---|---|---|---|---|
| | | Developed countries | Developing countries | | |
| 2003 | Current | 0.64 | 0.43 | 3 | 16.24 |
| 2020 | Current | 0.64 | 0.43 | 3.6 | 19.4 |
| | Affluent | 0.64 | 0.64 | 4.9 | 26.7 |
| 2050 | Current | 0.64 | 0.43 | 4.2 | 22.8 |
| | Affluent | 0.64 | 0.64 | 5.9 | 31.8 |

*Source: Smit et al. [107].*

**Table 2.**
*Global anticipated phosphorus consumption.*

when grown on additional marginal land with low P-status. From 1961 to 2013, total global P-consumption has increased five folds, to 31 million metric tons [106]. For 2050, with a global population of 9 billion, P-demand would rise by 40 and 96% for current and affluent diets, respectively [107]. **Table 2** shows the anticipated global P-consumption in developed and developing countries.

In 2020, China was the country with the largest consumption of phosphate fertilizers, with more than 11 million metric tons, which was followed by India and Brazil, with 8.98 and 6.04 million metric tons, respectively [108]. That year, ammonium phosphate was the most consumed phosphate fertilizer (over 25 million metric tons of $P_2O_5$) worldwide, followed by complex nitrogen, phosphorus, potassium (NPK) fertilizers (over 12 million metric tons of $P_2O_5$), as reported by Statista [109]. World consumption of $P_2O_5$ contained in fertilizer products increased by 7% in crop year 2021 compared with that in crop year 2020 [110]. The increases in world consumption and trade were driven by high crop prices, increased planted crop areas, and increased crop exports. This was a continuation of the trend that began late in 2020, as markets rebounded from poor weather conditions in the growing season. South America and Asia were leading regions of growth in consumption of phosphate fertilizer in terms of percentage increase over that in 2020. With the continuous rise in P-consumption, it is necessary to develop a system approach for addressing P supply, demand, and loss.

## 10. Management strategies for phosphorus to protect the environment

### 10.1 Phosphorus in the soil

Apatite and other primary minerals are the principal sources of P in the environment (calcium phosphate). Many rocks and soils include phosphate-bearing minerals. It takes time for these minerals to break down and release phosphorus into the soil. P is a highly reactive substance in the environment. This orthophosphate exists in a variety of different phosphate forms in solution, depending on the acidity. In the absence of rapid uptake by plants or soil micro-organisms, orthophosphate is likely to recombine with other soil chemicals (e.g., calcium, iron, aluminum, and manganese). As a result, P has long been regarded as the most important agricultural nutrient.

## 10.2 Phosphorus in the agriculture

Commercial, inorganic, phosphate fertilizers have been used to correct soil P-deficiency over the past 50 years. Plants can benefit from phosphate fertilizers because they are made by removing P from phosphate-rich rocks and making them more accessible. P has built up in agricultural soils due to massive volumes of these fertilizers over many decades. Crop yields are not adversely affected by soils with high P-reserves, except for the availability of micronutrients such as zinc. Before the 1970s, the practice of using high-phosphate fertilizers to increase soil phosphorus reserves was encouraged, and it was equated to "saving money in the bank." According to conventional belief, having high soil P-levels was equivalent to having significant cash in the bank. As a result of growing worries about the association between high soil P levels and the harm to water quality posed by P-rich soil particles entering the water through runoff, this concept has come under scrutiny in recent years.

## 10.3 Controlling runoff in case of agricultural phosphorus

Phosphate fertilizers, plant waste, manure, and agricultural effluent should not be directly applied to surface waters to prevent point source P-contamination. Pollution caused by nonpoint P-sources such as agricultural fields can be reduced by regulating the amount and kind of runoff. To avoid P pollution from surface waters, areas with steep slope and highly erodible soils close to surface waters should be avoided. A site's risk of becoming a prospective polluter can be considerably affected by management approaches. Sediment can be reduced significantly with the use of cover crops and buffer strips. Another factor that can enhance the risk of contamination is a lack of soil cover or barriers between a farm and its water.

Generally, water erosion happens on slopes, and the intensity of the decline increases with the slope's degree. Using no-till farming can reduce soil erosion. Additionally, crop residues can be left on the soil surface after harvest, and winter cover crops can be planted in the fall. Reduced water erosion can be achieved through the use of contoured tillage. Rather than going up and down the slope, plows are used to plow across it. The geometry of some fields makes this technique ineffective. To decrease water erosion, terraces can also be built. Soil erosion can be reduced by leaving agricultural remains on the field after harvesting. Leaving corn stalks on the field after harvest is one example of how crop residues can protect the soil from eroding throughout the winter. The residues protect the soil surface until the land is plowed, reducing the time it is exposed to the weather.

Protecting the soil from erosion during noncrop months is done by planting cover crops, such as grasses, legumes, or small grains. Once the main crop is planted, they remain in the field. Phosphorus levels in rivers and streams are reduced by cover crops, which reduce erosion. Cover crops can also take up P and other nutrients while other crops are not growing. Wildlife can also benefit from cover crops, which give food and shelter.

## 10.4 Management of fertilizer phosphorus sources

In agricultural runoff, the most prevalent sources of P are manure and compost, as well as mineral fertilizers [111]. Practices that can reduce the risk of agricultural land contamination with P include limiting the amount, timing, and method of application of P-fertilizer sources. There is a strong correlation between soil test P-levels (especially in newly fertilized plots, when the amount of soluble P in the fertilizer source is predominantly responsible for the dissolved P in surface runoff), with P-rate, and the timing of application, according to Mullins et al. [112]. Runoff water from fields getting broadcast P-fertilizer was shown to have a higher concentration of dissolved P than runoff water from locations where comparable quantities of P were absorbed 5 cm below the soil surface. All factors that enhance the danger of fertilizer loss, such as high fertilizer application rates and high rainfall regions, can be mitigated [113].

Additionally, some regions (e.g., areas with a lot of rainfall) should use less water-soluble fertilizers to decrease the transfer of P in runoff water [114]. When limiting incidental P-losses from fertilizer applications, scheduling the fertilizer P applications to correspond with dry weather is critical, especially in places with a lot of rainfall or regular irrigation [113]. Avoiding solid storms, which account for a significant portion of annual runoff P-loss, would prevent runoff P-loss. It has been found that delaying the application of P-nutrition sources until after a rainfall or runoff event reduces

| Management strategies | Agroecosystems | Effect on phosphorus transfer/losses | Causes/mechanisms | References |
|---|---|---|---|---|
| Soil application of organic wastes like pig slurry and cattle slurry | Maize under no-tillage systems, subtropical environment | Increased P transfer by surface runoff | Addition of higher amounts of P in soil and the doses of organic wastes were fixed according to N requirement of crops | Lourenzi et al. [76] |
| Selection of crop genotypes with shallow root growth | Common bean under low-input systems | Reduced P-losses in erosion (sediment) | Greater P-uptake and shoot growth | Henry et al. [121] |
| Application of NPK fertilizers and organic amendments like farmyard manure, green manure, and paddy straw | Rice-wheat cropping systems | Reduced P runoff | Protection of aggregated P by consolidation of micro aggregates into macro aggregates | Mitran et al. [122] |
| Reductions in fall-applied N and P-fertilizer inputs | Winter wheat-maize systems | Reduced P-transport in agricultural watersheds | Decreased winter wheat and increased corn production | Lerch et al. [123] |
| Reduced tillage and green manure | Wheat under Mediterranean rainfed systems | Reductions in P-losses 64% | Low mobilization of soil particles | Martínez-Mena et al. [124] |
| Low application of poultry manure combined with high N/P nutrient sources, high N fertilizers, and N-fixation from cover crops | Grain-tomato rotations | Low P leaching risks | Increased P sorption | Maltais-Landry et al. [125] |
| Magnesium-salt-coated biochar | Cereals under organic systems | Reduced P leaching | P-adsorption capacity of biochar layer | Riddle et al. [126] |

| Management strategies | Agroecosystems | Effect on phosphorus transfer/losses | Causes/mechanisms | References |
|---|---|---|---|---|
| No-tillage and unincorporated manure | Maize | Reduced P loads in runoff | Increased infiltration and decreased sediment losses | Bundy et al. [127] |
| Chicken manure biochar | Wheat | Decreased P leaching | Increased plant P-uptake due to an increase in soil pH and mycorrhizal colonization | Madiba et al. [128] |
| Gypsum | — | Decreased P leaching | Increased calcium content in soil | Favaretto et al. [129] |

**Table 3.**
*Selected reports on the management of phosphorus in agroecosystems.*

the amount of P that makes its way into runoff [115]. When it comes to runoff, the number of rainstorms following application can affect the concentration of nutrients more than the amount of runoff or rainfall that occurs each year, according to [114].

When managing P for environmental protection, the ultimate goal is to avoid the transportation of P-sources. According to this perspective, preventive and intercepting procedures have been developed by Sharpley et al. [116] to reduce P-transport. There are two distinct approaches to dealing with P pollution: preventive measures like cover crops and interception technology like buffer strips meant to remove P from the landscape [117]. In addition to conservation tillage and crop-residue management measures, buffer strips and terracing have been advocated to reduce P-movement through erosion and runoff. Cover crops, grassed waterways, and the development of riparian zones have also been proposed [118–120]. In **Table 3**, we have presented different management strategies that can reduce the P-losses from agricultural soils.

## 11. Conclusion and future perspectives

In conclusion, P deficiency in agricultural fields due to high soil fixation and limited nonrenewable P-stock are the major concerns for researchers globally. The active involvement of P in several physiological and biochemical functions has been well-documented in various studies. Limited P-availability in soil affects overall plant growth due to the decrease in P-uptake by plant roots. Application of P-containing fertilizers supplements the P-demand, but random application causes contamination of surface bodies. Applying P-fertilizers considering the soil test value and PUE can contribute to the sustainable management of P and mitigate environmental contaminations. Future research trials should focus on improving the understanding of P-uptake, -utilization, and -transport mechanisms under low P-environment. Further, extensive research is required in the field of root biology, along with identifying and enhancing gene expression for improved P-acquisition and use efficiencies.

## Author details

Tahsina Sharmin Hoque[1], Deepranjan Sarkar[2], Rahul Datta[3], Mohammad Golam Kibria[1], Rafi Ullah[4], Nazeer Ahmed[4], Mohammad Anwar Hossain[5*], Asim Masood[6] and Naser A. Anjum[6*]

1 Department of Soil Science, Bangladesh Agricultural University, Mymensingh, Bangladesh

2 Department of Agriculture, Integral Institute of Agricultural Science and Technology, Integral University, Lucknow, India

3 Department of Geology and Pedology, Faculty of Forestry and Wood Technology, Mendel University in Brno, Brno, Czech Republic

4 Department of Agriculture, University of Swabi, Anbar, Swabi-Khyber Pakhtunkhwa Pakistan

5 Department of Genetics and Plant Breeding, Bangladesh Agricultural University, Mymensingh, Bangladesh

6 Department of Botany, Aligarh Muslim University, Aligarh, India

*Address all correspondence to: anwargpb@bau.edu.bd and naanjum1@myamu.ac.in

## References

[1] Roberts TL, Johnston AE. Phosphorus use efficiency and management in agriculture. Resources, Conservation and Recycling. 2015;**105**:275-281

[2] Hinsinger P. Bioavailability of soil inorganic P in the rhizosphere as affected by root-induced chemical changes: A review. Plant and Soil. 2001;**237**(2):173-195

[3] Holford ICR. Soil phosphorus: Its measurement, and its uptake by plants. Soil Research. 1997;**35**(2):227-239

[4] Steen I. Phosphorus availability in the 21st century: Management of a non-renewable resource. Phosphorus and Potassium. 1998;**217**:25-31

[5] Cordell D, Drangert JO, White S. The story of phosphorus: Global food security and food for thought. Global Environmental Change. 2009;**19**: 292-305

[6] Becker C, Boersma M. Differential effects of phosphorus and fatty acids on *Daphnia magna* growth and reproduction. Limnology and Oceanography. 2005;**50**(1):388-397

[7] Menary RC, Staden JV. Effect of phosphorus nutrition and cytokinins on flowering in the tomato, *Lycopersicon esculentum* mill. Functional Plant Biology. 1976;**3**(2):201-205

[8] Sawan ZM, Fahmy AH, Yousef SE. Effect of potassium, zinc and phosphorus on seed yield, seed viability and seedling vigor of cotton (*Gossypium barbadense* L.). Archives of Agronomy and Soil Science. 2011;**57**(1):75-90

[9] Neumann G, Massonneau A, Martinoia E, Römheld V. Physiological adaptations to phosphorus deficiency during proteoid root development in white lupin. Planta. 1999;**208**(3):373-382

[10] Chan C, Liao YY, Chiou TJ. The impact of phosphorus on plant immunity. Plant and Cell Physiology. 2021;**62**(4):582-589

[11] Fredeen AL, Raab TK, Rao IM, Terry N. Effects of phosphorus nutrition on photosynthesis in *Glycine max* (L.) Merr. Planta. 1990;**181**(3):399-405

[12] Lopez-Bucio J, Cruz-Ramirez A, Herrera-Estrella L. The role of nutrient availability in regulating root architecture. Current Opinion in Plant Biology. 2003;**6**(3):280-287

[13] Sanchez PA. Phosphorus. In: Properties and Management of Soils in the Tropics. 2nd ed. Cambridge: Cambridge University Press; 2019. pp. 370-414

[14] Shen J, Yuan L, Zhang J, Li H, Bai Z, Chen X, et al. Phosphorus dynamics: From soil to plant. Plant Physiology. 2011;**156**:997-1005

[15] Weaver DM, Summers RN. Nutrient status (phosphorus). In: Report Card on Sustainable Natural Resource Use in Agriculture. Western Australia: Department of Agriculture and Food; 2013. pp. 134-143

[16] Mahler RL. Nutrients Plants Require for Growth. Moscow: College of Agricultural and Life Science, University of Idaho; CIS 1124. 2004

[17] Alori ET, Glick BR, Babalola OO. Microbial phosphorus solubilization and its potential for use in sustainable agriculture. Frontiers in Microbiology. 2017;**8**:971

[18] McLachlan KD, Elliot DE, De Marco DG, Garran JH. Leaf acid phosphatase isozymes in the diagnosis of phosphorus status in field-grown wheat. Australian Journal of Agricultural Research. 1987;**38**:1-13

[19] Föhse D, Claasen N, Jungk A. Phosphorus efficiency of plants: I. External and internal P requirement and P uptake efficiency of different plant species. Plant and Soil. 1988;**110**:101-109

[20] Rodríguez D, Andrade FH, Goudriaan J. Does assimilate supply limit leaf expansion in wheat grown in the field under low phosphorus availability? Field Crops Research. 2000;**67**:227-238

[21] Takenawa T, Itoh T. Phosphoinositides, key molecules for regulation of actin cytoskeletal organization and membrane traffic from the plasma membrane. Biochimica et Biophysica Acta. 2001;**1533**:190-206

[22] Yin HL, Janmey PA. Phosphoinositide regulation of the actin cytoskeleton. Annual Review of Physiology. 2003;**65**:761-789

[23] Xue H-W, Chen X, Mei Y. Function and regulation of phospholipid signalling in plants. Biochemical Journal. 2009;**421**:145-156

[24] Downes CP, Gray A, Lucocq JM. Probing phosphoinositide functions in signaling and membrane trafficking. Trends in Cell Biology. 2005;**15**(5):259-268

[25] Malhotra H, Vandana SS, Pande R. Phosphorus nutrition: Plant growth in response to deficiency and excess. In: Hasanuzzaman M, Fujita M, Nahar H, Oku K, Hawrylak-Nowak B, editors. Plant Nutrients and Abiotic Stress Tolerance. Singapore: Springer; 2018. pp. 171-190

[26] Rychter AM, Rao RM. Role of phosphorus in photosynthetic carbon metabolism. In: Pessarakali M, editor. Handbook of Photosynthesis. 2nd ed. Boca Raton: CRC Press; 2005. pp. 1-27

[27] Sondergaard TE, Schulz A, Palmgren MG. Energization of transport processes in plants. Roles of the plasma membrane H1-ATPase. Plant Physiology. 2004;**136**:2475-2482

[28] Parets-Soler A, Pardo JM, Serrano R. Immunocyto localization of plasma membrane H1-ATPase. Plant Physiology. 1990;**93**:1654-1658

[29] Jahn T, Baluska F, Michalke W, Harper JF, Volkmann D. A membrane H+-ATPase in the root apex: Evidence for strong expression in xylem parenchyma and asymmetric localization within cortical and epidermal cells. PhysiologiaPlantarum. 1998;**104**:311-316

[30] Zhang LY, Peng YB, Pelleschi-T ravier S, Fan Y, Lu YF, Lu YM, et al. Evidence for apoplasmic phloem unloading in developing apple fruit. Plant Physiology. 2004;**135**:574-586

[31] Jakobsen I. The role of phosphorus in nitrogen fixation by young pea plants (*Pisum sativum*). Plant Physiology. 1985;**64**:190-196

[32] Sa TM, Israel DW. Energy status and functioning of phosphorus-deficient soybean nodules. Plant Physiology. 1991;**97**:928-935

[33] Bolland M, Riley M, Thomson B, Paynter B, Baker M. Seed phosphorus: Its effect on plant production. Journal of the Department of Agriculture, Western Australia, Series 4. 1990;**31**(1):20-22

[34] Rose TJ, Pariasca-Tanaka J, Rose MT, Mori A, Wissuwa M. Seeds of doubt: Re-assessing the impact of grain P

concentrations on seedling vigor. Journal of Plant Nutrition and Soil Science. 2012;**175**:799-804

[35] Pariasca-Tanaka J, Vandamme E, Mori A, Segda Z, Saito K, Rose TJ, et al. Does reducing seed-P concentrations affect seedling vigor and grain yield of rice? Plant and Soil. 2015;**392**:253-266

[36] Rodriguez D, Zubillaga MM, Ploschuk EL, Keltjens WG, Goudriaan J, Lavado RS. Leaf area expansion and assimilate production in sunflower (*Helianthus annuus* L.) growing under low phosphorus conditions. Plant and Soil. 1998;**202**:133-147

[37] Halsted M, Lynch J. Phosphorus responses of C-3 and C-4 species. Journal of Experimental Botany. 1996;**47**:497-505

[38] Gniazdowska A, Mikulska M, Rychter AM. Growth, nitrate uptake and respiration rate in bean roots under phosphate deficiency. Biologia Plantarum. 1998;**41**:217-226

[39] Chen R, Song S, Li X, Liu H, Huang D. Phosphorus deficiency restricts plant growth but induces pigment formation in the flower stalk of Chinese kale. Horticulture, Environment and Biotechnology. 2013;**54**:243-248

[40] White PJ, Veneklaas EJ. Nature and nurture: The importance of seed phosphorus content. Plant and Soil. 2012;**357**:1-8

[41] Dordas C. Role of nutrients in controlling plant diseases in sustainable agriculture. A review. Agronomy for Sustainable Development. 2008;**28**: 33-46

[42] Tiwari KN. Phosphorus. In: Fundamentals of Soil Science. Revised Edition. New Delhi: Indian Society of Soil Science; 2012. pp. 413-430

[43] Emadi M, Baghernejad M, Emadi M, Fathi H, Saffari M. Phosphorus forms and behaviors in selected heavily fertilized soils. Archives of Agronomy and Soil Science. 2009;**55**(6):579-595

[44] Soremi AO, Adetunji MT, Azeez JO, Adejuyigbe CO, Bodunde JG. Speciation and dynamics of phosphorus in some organically amended soils of southwestern Nigeria. Chemical Speciation & Bioavailability. 2017;**29**(1):42-53

[45] Frossard E, Brossard M, Hedley MJ, Metherell A. Reactions controlling the cycling of P in soils. In: Tiessen H, editor. Phosphorus in the Global Environment. New York, NY, USA: John Wiley & Sons; 1995. pp. 107-137

[46] Havlin JL, Beaton JD, Tisdale SL, Nelson WL. Soil Fertility and Fertilizers. 7th ed. India: Pearson Education Inc.; 2012

[47] McLaren TI, Smernik RJ, McLaughlin MJ, McBeath TM, Kirby JK, Simpson RJ, et al. Complex forms of soil organic phosphorus–A major component of soil phosphorus. Environmental Science & Technology. 2015;**49**(22):13238-13245

[48] García-Albacete M, Martín A, Cartagena MC. Fractionation of phosphorus biowastes: Characterisation and environmental risk. Waste Management. 2012;**32**(6):1061-1068

[49] Requejo MI, Eichler-Löbermann B. Organic and inorganic phosphorus forms in soil as affected by long-term application of organic amendments. Nutrient Cycling in Agroecosystems. 2014;**100**(2):245-255

[50] Sarkar D, Rakesh S, Sinha AK, Mukhopadhyay P. Forms of phosphorus in some acidic Entisols of subtropical eastern India. International Journal of Plant and SoilScience. 2017;**19**(3):1-9

[51] Renneson M, Dufey J, Bock L, Colinet G. Effects of parental material and land use on soil phosphorus forms in southern Belgium. In: Proceedings of the 19th World Congress of Soil Science; Soil Solutions for a Changing World; 1-6 August 2010; Brisbane, Australia. Symposium 4.3.1 Impacts of Land Use Change in Unsustainable Ecosystems; 2010. pp. 40-43

[52] Sarkar D, Rakshit A, Al-Turki AI, Sayyed RZ, Datta R. Connecting bio-priming approach with integrated nutrient management for improved nutrient use efficiency in crop species. Agriculture. 2021;**11**(4):372

[53] Balemi T, Negisho K. Management of soil phosphorus and plant adaptation mechanisms to phosphorus stress for sustainable crop production: A review. Journal of Soil Science and Plant Nutrition. 2012;**12**(3):547-562

[54] Penn CJ, Camberato JJ. A critical review on soil chemical processes that control how soil pH affects phosphorus availability to plants. Agriculture. 2019;**9**(6):120

[55] Sukitprapanon TS, Jantamenchai M, Tulaphitak D, Prakongkep N, Gilkes RJ, Vityakon P. Influence of application of organic residues of different biochemical quality on phosphorus fractions in a tropical Sandy soil. Agronomy. 2021;**11**(2):248

[56] Ghosh D, Mandal M, Pattanayak SK. Long term effect of integrated nutrient management on dynamics of phosphorous in an acid Inceptisols of tropical India. Communications in Soil Science and Plant Analysis. 2021;**52**(19):1-15

[57] Jalali M, SajadiTabar S. Chemical fractionation of phosphorus in calcareous soils of Hamedan, western Iran under different land use. Journal of Plant Nutrition and Soil Science. 2011;**174**(4):523-531

[58] Farrell M, Macdonald LM, Butler G, Chirino-Valle I, Condron LM. Biochar and fertiliser applications influence phosphorus fractionation and wheat yield. Biology and Fertility of Soils. 2014;**50**(1):169-178

[59] Datta SC. A re-look on P supplying capacity of soil through a mechanistic approach. Journal of the Indian Society of Soil Science. 2016;**64**:S115-S216

[60] Costa MG, Gama-Rodrigues AC, Gonçalves JLDM, Gama-Rodrigues EF, Sales MVDS, Aleixo S. Labile and non-labile fractions of phosphorus and its transformations in soil under eucalyptus plantations, Brazil. Forests. 2016;**7**(1):15

[61] Tarafdar JC. Phosphorus – A bright future ahead. Journal of Indian Society of Soil Science. 2013;**61**:S29-S37

[62] Bünemann EK, Oberson A, Liebisch F, Keller F, Annaheim KE, Huguenin-Elie O, et al. Rapid microbial phosphorus immobilization dominates gross phosphorus fluxes in a grassland soil with low inorganic phosphorus availability. Soil Biology and Biochemistry. 2012;**51**:84-95

[63] Raven KP, Hossner LR. Sorption and desorption quantity-intensity parameters related to plant-available soil phosphorus. Soil Science Society of America Journal. 1994;**58**(2):405-410

[64] Holford ICR. Greenhouse evaluation of four phosphorus soil tests in relation to phosphate buffering and labile phosphate in soils. Soil Science Society of America Journal. 1980;**44**(3):555-559

[65] Shirvani M, Shariatmadari H, Kalbasi M. Phosphorus buffering

capacity indices as related to soil properties and plant uptake. Journal of Plant Nutrition. 2005;**28**(3):537-550

[66] McDowell RW, Sharpley AN. Phosphorus solubility and release kinetics as a function of soil test P concentration. Geoderma. 2003;**112**(1-2):143-154

[67] Verheyen D, Van Gaelen N, Ronchi B, Batelaan O, Struyf E, Govers G, et al. Dissolved phosphorus transport from soil to surface water in catchments with different land use. Ambio. 2015;**44**(2):228-240

[68] Haygarth PM, Heathwaite AL, Jarvis SC, Harrod TR. Hydrological factors for phosphorus transfer from agricultural soils. Advances in Agronomy. 2000;**69**:153-178

[69] Bhandari SC. Soil phosphorus management and environmental factors. Journal of the Indian Society of Soil Science. 2013;**61**:S49-S58

[70] Shigaki F, Sharpley A, Prochnow LI. Rainfall intensity and phosphorus source effects on phosphorus transport in surface runoff from soil trays. Science of the Total Environment. 2007;**373**(1):334-343

[71] Sharpley A, Moyer B. Phosphorus forms in manure and compost and their release during simulated rainfall. Journal of Environmental Quality. 2000;**29**:1462-1469

[72] Giles CD, Cade-Menun BJ, Liu CW, Hill JE. The short-term transport and transformation of phosphorus species in a saturated soil following poultry manure amendment and leaching. Geoderma. 2015;**257**:134-141

[73] Molina M, Aburto F, Calderón R, Cazanga M, Escudey M. Trace element composition of selected fertilizers used in Chile: Phosphorus fertilizers as a source of long-term soil contamination. Soil and Sediment Contamination. 2009;**18**(4):497-511

[74] Tirado R, Allsopp M. Phosphorus in agriculture: Problems and solutions. Greenpeace Research Laboratories Technical Report. Amsterdam, Netherlands: Greenpeace International; 2012. pp. 3-30

[75] Sahu SK, Ajmal PY, Bhangare RC, Tiwari M, Pandit GG. Natural radioactivity assessment of a phosphate fertilizer plant area. Journal of Radiation Research and Applied Sciences. 2014;**7**(1):123-128

[76] Lourenzi CR, Ceretta CA, Ciancio NHR, Tiecher TL, da Silva LOS, De Conti L, et al. Forms of nitrogen and phosphorus transfer by runoff in soil under no-tillage with successive organic waste and mineral fertilizers applications. Agricultural Water Management. 2021;**248**:106779

[77] Withers PJ, Sylvester-Bradley R, Jones DL, Healey JR, Talboys PJ. Feed the crop not the soil: Rethinking phosphorus management in the food chain. Environmental Science and Technology. 2014;**48**(12):6523-6530

[78] Hammond JP, Broadley MR, White PJ, King GJ, Bowen HC, Hayden R, et al. Shoot yield drives phosphorus use efficiency in *Brassica oleracea* and correlates with root architecture traits. Journal of Experimental Botany. 2009;**60**:1953-1968

[79] FAO. Fertilizer and Plant Nutrition Bulletin. Rome: Food and Agriculture Organization of the United Nations (FAO); 2008

[80] Zhang H, Hardy DH, Mylavarapu R, Wang JJ. Mehlich-3 Soil Test Methods

from the Southeastern United States. Sikora FJ. and Moore K. (Eds.), Southern Cooperative Series Bulletin 419. Athens, GA: University of Georgia; 2014. pp. 101-110

[81] Sharpley A, Kleinman P, Baffaut C, Beegle D, Bolster C, Collick A, et al. Evaluation of phosphorus site assessment tools: Lessons from the USA. Journal of Environmental Quality. 2017;**46**(6):1250-1256

[82] Liuzza LM, Bush EW, Tubana BS, Gaston LA. Determining nutrient recommendations for agricultural crops based on soil and plant tissue analyzes between different analytical laboratories. Communications in Soil Science and Plant Analysis. 2020;**51**(3):392-402

[83] Nesse P, Grava J. Correlation of several tests for phosphorus with resin extractable phosphorus on alkaline soils. In: A Report on Field Research in Soils. Vol. 2. Univ. of Minnesota Agric. Exp. Stn. Misc. Publ.; 1986. pp. 218-220

[84] Olsen SR, Cole CV, Watanabe FS, Dean LA. Estimation of Available Phosphorus in Soils by Extraction with Sodium Bicarbonate. Washington, DC: USDA Circ. 939. U.S. Gov. Print. Office; 1954

[85] Blanchar RW, Caldwell AC. Phosphorus uptake by plants and readily extractable phosphorus in soils. Agronomy Journal. 1964;**56**(2):218-221

[86] Fixen PE, Carson PL. Relationship between soil test and small grain response to P fertilization in field experiments. Agronomy Journal. 1978;**70**(5):838-844

[87] Randall GW, Evans SD, Nelson WW. High phosphorus and potassium rates in a corn-soybean rotation. In: A Report on Field Research in Soils, Univ. of Minnesota Agric. Exp. Stn. Misc. Publ. 2. 1986. pp. 153, 153-164, 164

[88] John MK, vanRyswyk AL, Mason JL. Effect of soil order, pH, texture and organic matter on the correlation between phosphorus in alfalfa and soil test values. Canadian Journal of Soil Science. 1967;**47**:157-161

[89] Thompson LF, Pratt PF. Solubility of phosphorus in chemical extractants as indexes to available phosphorus in Ohio soils. Soil Science Society of America Journal. 1954;**18**:467-470

[90] Abdu N. Soil-phosphorus extraction methodologies: A review. African Journal of Agricultural Research. 2006;**1**:159-161

[91] Syers JK, Johnston AE, Curtin D. Efficiency of soil and fertilizer phosphorus use: Reconciling changing concepts of soil phosphorus behaviour with agronomic information. In: FAO Fertilizer and Plant Nutrition Bulletin. Rome: Food and Agriculture Organization of the United Nations; 2008. p. 11

[92] Negassa W, Leinweber P. How does the Hedley sequential phosphorus fractionation reflect impacts of land use and management on soil phosphorus: A review. Journal of Plant Nutrition and Soil Science. 2009;**172**:305-325

[93] Sugihara S, Funakawa S, Nishigaki M, Kilasara M, Kosaki T. Dynamics of fractionated P and P budget in soil under different land management in two Tanzanian croplands with contrasting soil textures. Agriculture, Ecosystems and Environment. 2012;**162**:101-107

[94] Haynes RJ. Effects of lime on phosphate availability in acid soil: A critical review. Plant and Soil. 1982;**68**:289-308

[95] Sumner ME, Farina MPW. Phosphorus interactions with other nutrients and lime in field cropping systems. Advances in Soil Science. 1986;**5**:201-236

[96] Holford ICR, Scheirtzer BE, Crocker GJ. Long-term effects of lime on soil phosphorus solubility and sorption in eight soils. Australian Journal of Soil Science. 1994;**32**:795-803

[97] Curtin D, Syers JK. Lime-induced changes in indices of soil phosphate availability. Soil Science Society of America Journal. 2001;**65**:147-152

[98] Johnston AE, Poulton PR. The role of phosphorus in crop production and soil fertility: 150 years of field experiments at Rothamsted. In: Schultz JJ, editor. Phosphate Fertilizers and the Environment. Muscle Shoals, USA: International Fertilizer Development Center; 1992. pp. 45-64

[99] Owino-Gerroh C, Gasch GJ. Effect of silicon on low pH soil phosphorus sorption and on uptake and on growth of maize. Communications in Soil Science and Plant Analysis. 2004;**15&16**:2369-2378

[100] White MJ, Storm DE, Busteed PR, Stoodley SH, Philips SJ. Evaluating non-point source critical source area contributions at the watershed scale. Journal of Environmental Quality. 2009;**38**(4):1654-1663

[101] Radcliff DE, Cabrera ML. Suggestions to improve modeling of phosphorus. In: Radcliffe DE, Cabrera ML, editors. Modeling Phosphorus in the Environment. Boca Raton, Florida: CRC Press; 2007. pp. 405-415

[102] Sharpley AN. Modeling phosphorus movement from agriculture to surface waters. In: Radcliffe DE, Cabrera ML, editors. Modeling Phosphorus in the Environment. Boca Raton, Florida: CRC Press; 2007. pp. 3-19

[103] Lemunyon JL, Gilbert RG. The concept and need for a phosphorus assessment tool. Journal of Production Agriculture. 1993;**6**(4):483-486

[104] Sharpley AN, Beegle D, Bolster C, Good L, Joern B, Ketterings Q, et al. Phosphorus indices: Why we need to take stock of how we are doing. Journal of Environmental Quality. 2012;**41**:1711-1719

[105] Nelson NO, Shober AL. Evaluation of phosphorus indices after twenty years of science. Journal of Environmental Quality. 2012;**41**:1703-1710

[106] Chen M, Graedel TEA. Half-century of global phosphorus flows, stocks, production, consumption, recycling and environmental impacts. Global Environmental Change. 2016;**36**:139-152

[107] Smit AL, Bindraban PS, Schröder JJ, Conijin JG, van der Meer HG. Phosphorus in Agriculture: Global Resources, Trends and Developments. Plant Research International: Wageningen; 2009. pp. 1-36

[108] Statista 2023. Consumption of phosphate fertilizer worldwide in 2020, by country

[109] Statista 2023. Consumption of phosphate fertilizer worldwide in 2020, by select product

[110] Jasinski SM. Phosphate Rock. Mineral Commodity Summaries. Reston, Virginia: U.S. Geological Survey; 2022. p. 125

[111] Sharpley AN, Chapra SC, Wedepohl R, Sims JT, Daniel TC, Reddy KR. Managing agricultural phosphorus for protection

of surface waters: Issues and options. Journal of Environmental Quality. 1994;**23**:437-451

[112] Mullins G, Joern B, Moore P. By-product phosphorus: Sources, characteristics, and management. In: Sims JT, Sharpley AN, editors. Phosphorus: Agriculture and the Environment. Vol. 46. Madison, Wisconsin, USA: American Society of Agronomy; 2005. pp. 829-879

[113] Withers PJA, Ulén B, Stamm C, Bechmann M. Incidental phosphorus losses—Are they significant and can they be predicted? Journal of Plant Nutrition and Soil Science. 2003;**166**:459-468

[114] Zaimes GN, Schultz RC. Phosphorus in Agricultural Watersheds. Ames, Iowa: Department of Forestry at Iowa State University; 2002

[115] Sharpley AN, Beegle D. Managing phosphorus for agriculture and the environment. Pennsylvania: Pennsylvania State University; 2001

[116] Sharpley AN, Foy B, Withers P. Practical and innovative measures for the control of agricultural phosphorus losses to water: An overview. Journal of Environmental Quality. 2000;**29**:1-9

[117] Sharpley AN. Agriculture and Phosphorus Management: The Chesapeake Bay. Boca Raton, Florida: Lewis Publishers; 2000

[118] Sharply AN, Withers PA. The environmentally sound management of agricultural phosphorus. Fertilizer Research. 1994;**39**(2):133-146

[119] Sims JT, Maguire RO, Leytem AB, Gartley KL, Pautler MC. Evaluation of Mehlich 3 as an Agri-environmental soil phosphorus test for the mid-Atlantic United States of America. Soil Science Society of America Journal. 2002;**66**:2016-2032

[120] Sims JT, Kleinman PJA. Managing agricultural phosphorus for environmental protection. In: Sims JT, Sharpley AN, editors. Phosphorus: Agriculture and the Environment. Vol. 46. Madison, Wisconsin, USA: American Society of Agronomy; 2005. pp. 1021-1068

[121] Henry A, Chaves NF, Kleinman PJ, Lynch JP. Will nutrient-efficient genotypes mine the soil? Effects of genetic differences in root architecture in common bean (*Phaseolus vulgaris* L.) on soil phosphorus depletion in a low-input agro-ecosystem in Central America. Field Crops Research. 2010;**115**(1):67-78

[122] Mitran T, Mani PK, Bandyopadhyay PK, Basak N. Effects of organic amendments on soil physical attributes and aggregate-associated phosphorus under long-term rice-wheat cropping. Pedosphere. 2018;**28**(5):823-832

[123] Lerch RN, Baffaut C, Kitchen NR, Sadler EJ. Long-term agroecosystem research in the Central Mississippi River basin: Dissolved nitrogen and phosphorus transport in a high-runoff-potential watershed. Journal of Environmental Quality. 2015;**44**(1):44-57

[124] Martínez-Mena M, Carrillo-López E, Boix-Fayos C, Almagro M, Franco NG, Díaz-Pereira E, et al. Long-term effectiveness of sustainable land management practices to control runoff, soil erosion, and nutrient loss and the role of rainfall intensity in Mediterranean rainfed agroecosystems. Catena. 2020;**187**:104352

[125] Maltais-Landry G, Scow K, Brennan E, Vitousek P. Long-term effects

of compost and cover crops on soil phosphorus in two California agroecosystems. Soil Science Society of America Journal. 2015;**79**(2):688-697

[126] Riddle M, Bergström L, Schmieder F, Lundberg D, Condron L, Cederlund H. Impact of biochar coated with magnesium (hydr) oxide on phosphorus leaching from organic and mineral soils. Journal of Soils and Sediments. 2019;**19**(4):1875-1889

[127] Bundy LG, Andraski TW, Powell JM. Management practice effects on phosphorus losses in runoff in corn production systems. Journal of Environmental Quality. 2001;**30**(5):1822-1828

[128] Madiba OF, Solaiman ZM, Carson JK, Murphy DV. Biochar increases availability and uptake of phosphorus to wheat under leaching conditions. Biology and Fertility of Soils. 2016;**52**(4):439-446

[129] Favaretto N, Norton LD, Johnston CT, Bigham J, Sperrin M. Nitrogen and phosphorus leaching as affected by gypsum amendment and exchangeable calcium and magnesium. Soil Science Society of America Journal. 2012;**76**(2):575-585

Chapter 5

# Regulation of Plant-Microbe Interactions in the Rhizosphere for Plant Growth and Metabolism: Role of Soil Phosphorus

*Anurag Yadav and Kusum Yadav*

## Abstract

Soil phosphorus (P) plays a crucial role in regulating plant-microbe interactions in the rhizosphere. This chapter provides an in-depth analysis of the role of P in plant physiology, growth, and its availability in soil. Phosphorus acquisition and uptake, its impact on plant metabolism, and its influence on plant growth and development are reviewed in this chapter. The mechanisms by which plants acquire P from the soil, including the production of phosphatase enzymes, secretion of organic acids, mycorrhizal symbiosis, changes in root morphology, phosphorus use efficiency, and mobilization and transporters, are also reviewed. The chapter also explores the impact of P on microbial communities in the rhizosphere and its role in plant-microbe interactions. Finally, the implications of P availability in the rhizosphere for sustainable agriculture and crop production are discussed, highlighting the potential for improving P availability in the soil to enhance agricultural productivity and environmental sustainability.

**Keywords:** soil phosphorus, plant-microbe interactions, phosphorus acquisition and uptake, rhizosphere microbial communities, sustainable agriculture and crop production

## 1. Introduction

The rhizosphere, the region of soil surrounding plant roots, is a dynamic environment where various interactions between plants and microorganisms occur [1]. Among these interactions, the exchange of nutrients is crucial for plant growth and development. Phosphorus (P) is one of the essential macronutrients for plants known to play a vital role in several physiological processes, including photosynthesis, energy transfer, and nucleic acid synthesis [2]. However, P-availability in the soil is often limited since it is highly reactive and readily forms insoluble compounds unavailable to plants [3]. Plants have evolved various strategies to acquire P from soil to overcome this limitation. These mechanisms involve the secretion of organic acids, enzymes, and other compounds that can solubilize and mineralize P [4] through associations with beneficial microorganisms such as mycorrhizal fungi and rhizobacteria [5].

The interactions between plants and microorganisms in the rhizosphere play a crucial role in regulating P-availability [3]. Microorganisms can contribute to the solubilization and mineralization of P and make it available to plants [6]. Conversely, some microorganisms can immobilize or compete for P and reduce its plant availability [7]. Therefore, understanding the dynamics of plant-microbe interactions in the rhizosphere is essential for improving P-acquisition and enhancing plant growth and yield [8]. Moreover, sustainable agriculture and crop production practices require minimizing chemical fertilizers and enhancing the natural processes of nutrient cycling in soil [9]. Harnessing plant-microbe interactions to optimize P-uptake can reduce the environmental impact of agriculture while improving soil health and crop productivity [10]. Therefore, studying the role of P in plant-microbe interactions in the rhizosphere is critical for developing efficient and sustainable agricultural systems.

This chapter aims to: (i) provide an in-depth analysis of the role of P in plant physiology, growth, and its availability in soil; (ii) discusses P-acquisition and -uptake, its impact on plant metabolism and its influence on plant growth and development; (iii) examines the mechanisms underlying the soil-P acquisition in plants, and the production of phosphatase enzymes, secretion of organic acids, mycorrhizal symbiosis, changes in root morphology, P-use-efficiency, and mobilization and transporters; (iv) explores the impact of P on microbial communities in the rhizosphere and the role it plays in plant-microbe interactions; and (v) highlights the significant implications of P-availability in the rhizosphere for sustainable agriculture and crop production.

## 2. The role of phosphorus in plant physiology and growth

Among the most critical essential macronutrients, P is required by plants for their proper growth and development. It plays a crucial role in plants' physiological and biochemical processes, including photosynthesis, energy transfer, cell division, nucleic acid synthesis, and membrane transport. Therefore, understanding the role of P in plant physiology and growth is crucial for plant scientists, agronomists, and farmers to optimize crop production and improve the sustainability of agricultural systems.

### 2.1 Phosphorus acquisition and uptake

Phosphorus is an essential nutrient for plant growth and development, playing a crucial role in various metabolic processes such as ATP synthesis and nucleic acid formation [2]. Plants require large amounts of P, which is often the limiting factor in its growth [10]. Plants take it up as phosphate ions ($H_2PO_4$ and $HPO_4^{2-}$) found in the soil [11]. The availability of phosphate ions in the soil is affected by soil pH, temperature, moisture, and microbial activity [3]. However, P is often present in limited amounts in soil, mainly in insoluble phosphates [12]. Therefore, plants have developed various mechanisms to acquire and uptake P from the soil [13]. One of the strategies plants use to obtain P is through production of phosphatase enzymes (**Figure 1**). These enzymes can hydrolyze organic phosphates in the soil, releasing inorganic phosphate ions that the plant can take up. Several studies have shown that plants can increase the production of phosphatase enzymes in response to low soil-P levels [4, 14]. Plants can also secrete organic acids, such as citrate and malate, to solubilize insoluble P compounds in

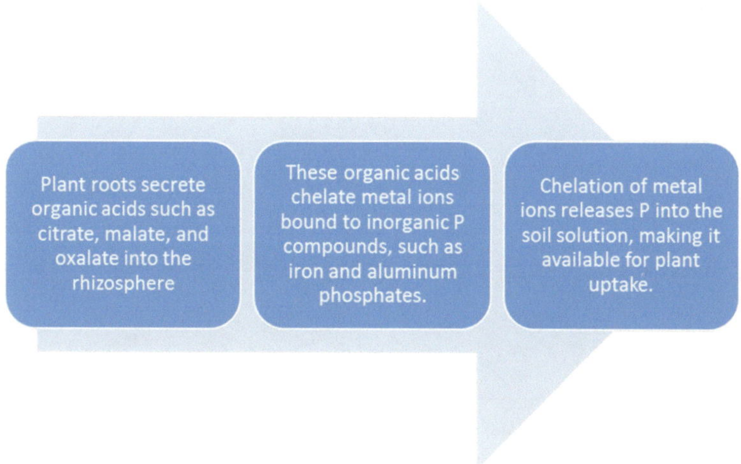


**Figure 1.**
*Representation of the major processes involved in phosphorus-aquision by plants involving organic acid secretion.*

the soil. These organic acids can chelate with metal ions, reducing their ability to bind with phosphates and making them more available for plant uptake [15].

Furthermore, many plants form symbiotic associations with mycorrhizal fungi, which can enhance P-uptake and -translocation in the plant. Mycorrhizal fungi can extend their hyphae into the soil, increasing the surface area for P-uptake. They can also release phosphatase enzymes and organic acids, increasing P-availability [16].

### 2.2 Phosphorus and plant metabolism

Phosphorus is vital in plant metabolism as an essential macronutrient for growth and development. According to Rao and Pessarakli [17], P is absorbed by plant roots as phosphate ions and then transported to various plant tissues and organs (**Figure 2**). Once inside the plant, P is involved in numerous metabolic pathways, including photosynthesis and respiration. In photosynthesis, P is a fundamental component of ATP, the primary energy source for plant cells. ATP is synthesized in the thylakoid membranes of chloroplasts through the phosphorylation of ADP (adenosine diphosphate) using energy from light [10]. P also regulates photosynthetic carbon metabolism by activating specific enzymes and proteins involved in the process. P is also crucial to nucleic acids, such as DNA and RNA. Nucleic acids are involved in gene expression and protein synthesis, essential for plant growth and development. In addition, P is involved in forming phospholipids, crucial components of cell membranes [18]. Phospholipids serve as precursors for synthesizing signaling molecules, such as inositol phosphates, which are involved in various cellular signaling pathways [19]. P also plays a crucial role in regulating various metabolic processes in plants. P regulates the synthesis and activity of various carbohydrate, lipid, and protein metabolism enzymes [20].

Understanding the role of P in plant metabolism is crucial for optimizing its availability in the soil and improving agricultural productivity and sustainability. Research into P and its plant functions can help develop strategies to enhance soil P levels, leading to healthier and more productive crops.

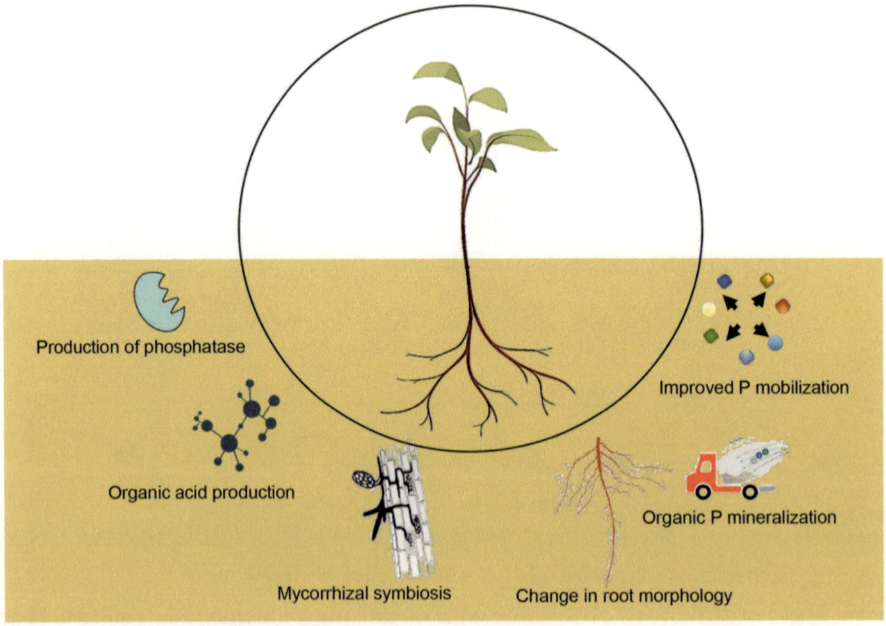


**Figure 2.**
*Representation of the major modes of phosphorus uptake in plants.*

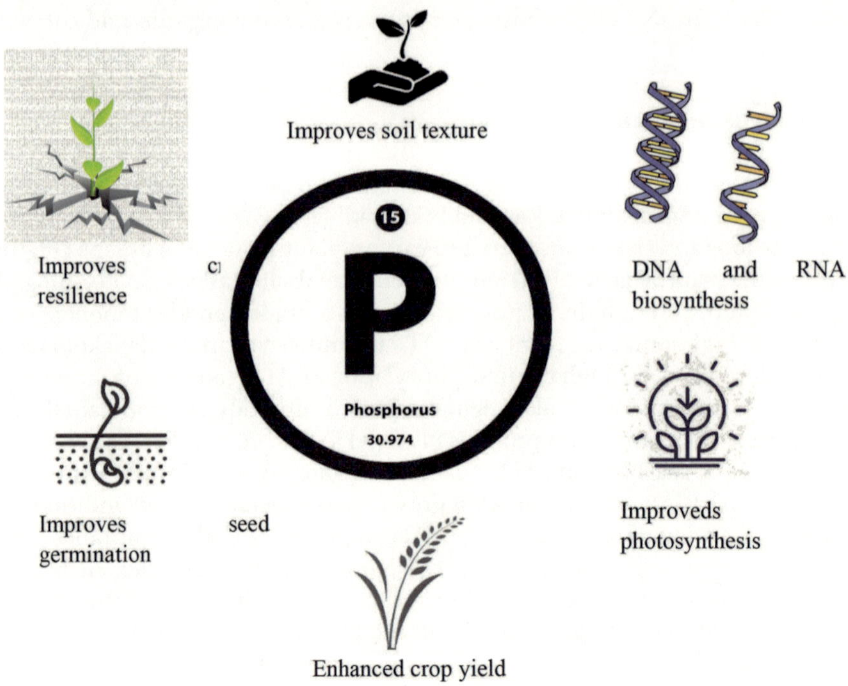


**Figure 3.**
*Representation of the major roles of phosphororus in plant growth, development and yield.*

## 2.3 Phosphorus and plant growth development

Phosphorus is essential for plant growth and development [21]. It is required for cell division and elongation, which are critical processes in plant growth [2]. P deficiency can lead to stunted growth, reduced seed production, and poor fruit quality in plants [11]. P also plays a vital role in root development, crucial for nutrient and water uptake from the soil [22]. In addition, P can improve plant resistance to environmental stresses, such as drought, heat, and cold, by regulating the expression of stress-related genes and enhancing the production of stress-related proteins [23] (**Figure 3**). **Table 1** lists the major phosphate-solubilizing microorganisms reported beneficial for plant growth.

| Microorganism | Test plant | Benefit(s) to plants | References |
|---|---|---|---|
| *Aspergillus niger* | Munbean | Improved P and N uptake, heat tolerance | [24] |
| *Aspergillus niger* | Wheat | increased the growth and phosphate uptake | [25] |
| *Aspergillus niger* | Rice | Rock phosphate solubilization, and improved growth | [26] |
| *Azospirillum brasilense* | Wheat | Increased yield | [27] |
| *Azotobacter chroococcum* | Wheat | Enhanced phosphorus uptake, symbiotic growth | [28, 29] |
| *Achromobacter piechaudii* | Rice | Enhanced phosphorus uptake, yield | [30] |
| *Bacillus subtilis* | Munbean | Improved P and N uptake, heat tolerance | [24] |
| *Bacillus circulans* | Munbean | Improved P and N uptake, heat tolerance | [24] |
| *B. subtilis* | Tomato | Improved yield | [31] |
| *Bacillus megaterium* | Wheat | Enhanced phosphorus uptake, increased growth | [6, 32, 33] |
| *Burkholderia cepacia* | Chickpea | Alleviated glyphosate induced toxicity | [34] |
| *Enterobacter ludwigii* | Rice | Improved biomass and chloropohyll content | [35] |
| *Glomus sp.* (mycorrhizal fungi) | Sorghum | Increased phosphorus uptake and biomass | [36] |
| *Penicillium bilaii* | Barley | Improved phosphorus uptake and yield | [37] |
| *Pantoea agglomerans* | Maize | Improved phosphorus availability and yield | [38, 39] |
| *Pseudomonas fluorescens* | Wheat | Increased growth and P levels | [40] |
| *P. fluorescens* | Tomato | plant growth and increased total root length, surface area and volume | [41] |
| *P. fluorescens* | Maize | Improved phosphorus availability, yield | [42] |
| *Serratia marcescens* | Maize | Improved biomass and yield | [43] |
| *Pseudomonas* sp. | Maize | Improved biomass and yield | [43] |
| *S. marcescens* | Wheat | Enhanced phosphorus uptake, increased growth | [44] |

| Microorganism | Test plant | Benefit(s) to plants | References |
|---|---|---|---|
| *Streptomyces sp.* | Rice | Improved plant growth by antagonizing phytopathogens | [45] |
| *Trichoderma viride* | Cucumber | Improved growth | [46] |
| *Trichoderma viride* | Soybean | Improved phosphorus uptake | [47] |

**Table 1.**
*Major phosphate solubilizing microorganisms reported beneficial for plant growth.*

## 3. Phosphorus availability in soil and mechanisms underlying its acquisition in plants

### 3.1 Production of phosphatase enzymes

Plants have developed a sophisticated network of mechanisms to cope with various environmental stresses, such as nutrient deficiency, drought, and high salinity. One crucial strategy plants employ is through the production of phosphatase enzymes, which play a crucial role in mobilizing P from various organic and inorganic sources in the soil [4]. The secretion of phosphatase enzymes by plant roots into the rhizosphere enables the breakdown of a wide range of organic P-based compounds, making inorganic P readily available for plant uptake [48, 49]. Under low P-conditions, plants induce the expression of genes encoding phosphatases, increasing enzyme activity in various tissues [50]. This upregulation of phosphatase activity helps plants to efficiently scavenge phosphate from the soil and maintain their growth and development. Moreover, mycorrhizal fungi can stimulate phosphatase activity, enhancing P uptake and translocation in plants [4].

The major phosphatases identified in plants include acid phosphatases, alkaline phosphatases, and purple acid phosphatases [4]. These enzymes are involved in the hydrolysis of different types of P-compounds, such as nucleotides, phospholipids, and phytate. Phosphatase enzymes have been shown to have multiple functions in plants, such as regulating iron homeostasis and improving the availability of minerals like iron, zinc, and calcium for plant uptake [4, 14].

### 3.2 Secretion of organic acids

Plants also secrete organic acids, such as citrate and malate, into the rhizosphere to solubilize inorganic P-compounds such as iron and aluminum phosphates [51]. These acids chelate metal ions bound to P, releasing P into the soil solution and making it available for plant uptake [52]. Through this mechanism plants cope with P deficiency and low availability in the soil [51]. The secretion of organic acids by plant roots facilitate the solubilization and uptake of essential nutrients, including P, iron, and aluminum [53]. Organic acids are low-molecular-weight compounds that chelate cations and dissolve insoluble mineral compounds, making them more available for plant uptake [51].

Plant roots, including citrate, malate, and oxalate, secrete several organic acids [15]. Citrate is the most commonly secreted organic acid, and is involved in the solubilization and uptake of various nutrients, such as iron, aluminum, and P [54]. On the other hand, malate is involved in the uptake of aluminum, while oxalate helps in the uptake of calcium and aluminum [15]. These organic acids are secreted into the rhizosphere by specialized cells in the root, known as root border cells, and root hairs [53].

Various factors, including nutrient deficiency, pH, and root exudates, regulate the secretion of organic acids in plants [55]. Under nutrient-deficient conditions, plants induce the expression of genes encoding enzymes involved in organic acid synthesis and transport, increasing their secretion [51]. Additionally, the pH of the rhizosphere can also influence organic acid secretion, with lower pH values generally leading to increased secretion [55].

Releasing organic acids by plant roots may also substantially affect the soil microbial ecology. Organic acids can act as a carbon source for soil microorganisms, promoting their growth and activity [56]. Additionally, the solubilization of nutrients by organic acids can also increase microbial activity and diversity in the rhizosphere. Several studies have reported on the secretion of organic acids by different plant species. For instance, a study by Palomo, Claassen [57] showed that maize plants secrete high citrate levels under P-deficient conditions, enhancing their P uptake. In another similar study soybean plants were reported to secrete malate and oxalate under aluminum stress conditions, which was argued to increase aluminum tolerance [58].

### 3.3 Mycorrhizal symbiosis

P, an essential nutrient for plant growth and development. However, it is often limited in soil due to low availability and solubility. The mycorrhizal symbiosis enables plants to overcome P-limitation by accessing P from a larger soil volume and increasing P-uptake efficiency by forming highly branched arbuscules. Mycorrhizal symbiosis is a mutually beneficial relationship between plants and fungi, where the plant provides the fungus with carbohydrates in exchange for nutrients, including P. The arbuscules provide a large surface area for P uptake and translocation from the soil to the plant.

Mycorrhizal fungi can also release enzymes and organic acids that help to solubilize P-compounds in soil, making them available for plant uptake. The release of organic acids by mycorrhizal fungi, such as citrate, malate, and oxalate, can increase the solubility of insoluble P compounds, such as calcium phosphates, by forming complexes with metal ions that bind to P, thus freeing it for plant uptake [59]. Additionally, mycorrhizal fungi can release enzymes such as acid phosphatase, which can hydrolyze organic P-compounds, such as phytate, and release inorganic P for plant take up [4].

### 3.4 Change in root morphology

One common strategy plants employ for increasing P-uptake is the production of longer and more branched roots, which can increase the surface area of the root system and improve the plant's ability to explore the soil for nutrients. For example, several studies have shown that plants grown under low P-conditions produce longer and more branched roots than those grown under high P-conditions [60, 61].

In addition to changes in root length and branching, plants can also alter the distribution of root hairs, which are small projections from the root surface that increase the surface area of the root system. Under low P-conditions, some plant species produce more root hairs per unit length of root than under high P conditions [62]. Plants can also alter the morphology of their root tips to improve P-uptake. For example, some plants produce cluster roots, which are highly branched structures that form at the tips of roots and increase the surface area of the root system. These structures are prevalent in plants that grow in soils with low P-availability, such as Proteaceae and Casuarinaceae species [63]. Another strategy that plants use to increase P-uptake is the production of exudates, which are organic compounds released by plant roots that can increase P-availability in soil. For example, some plants release organic acids that can solubilize mineral-bound P and make it available for uptake [4]. In addition to these morphological changes, some plant species have also evolved symbiotic relationships with mycorrhizal fungi, which can improve P-uptake by extending the root system and increasing the surface area of the root system [5].

### 3.5 Phosphorus use-efficiency

Plants can increase their P use-efficiency by adopting various strategies to optimize P-metabolism in the soil. One of the primary strategies is to enhance the uptake of P from the soil, which is achieved by developing an extensive root system that allows plants to explore a larger volume of soil, and the secretion of organic acids and enzymes that solubilize and release bound P in the soil [64]. Plants can increase their P use-efficiency by recycling and reusing internal P reserves. For example, plants can remobilize P from old leaves to new growth areas during leaf senescence, improving their ability to acquire P from the soil [65]. This way the plants recycles and reuses their own internal reserves of P. Plants can remobilize P from old leaves, stems, and other tissues to support new growth, reducing their reliance on external sources of P [66]. This process is especially important under low P availability conditions, where efficiently reallocating P from senescent tissues to new growth can improve plant fitness and productivity [67].

### 3.6 Phosphorus mobilization and transporters

Plants can mobilize and transport P within their tissues to improve their ability to acquire P from the soil. P can be mobilized within plant tissues by phosphatase enzymes, which break down organic P-compounds into inorganic forms that are more readily available for plant uptake. Plants also have specific transporters that can uptake inorganic P from the soil and transport it into their tissues [68].

Membrane proteins are called P-transporters to facilitate P transport across plants' plasma membranes and other intracellular membranes [69]. Several P-transporters exist in plants including the PHT1, PHT2, PHT3, PHO1 transporters [70]. The PHT1 transporter family is the most extensively studied and is involved in inorganic phosphate uptake from the soil [71]. This transporter family comprises 9 to 13 members in different plant species and is expressed in the root epidermis and cortex, where they play a crucial role in the uptake of inorganic phosphate from the soil [71]. They have been shown to have a high affinity for inorganic phosphate and can transport it against a concentration gradient [72]. PHT2 transporters are expressed in the plasma membrane of root hairs and are involved in phosphate uptake and translocation [54].

PHT3 transporters are localized in the chloroplast and are involved in phosphate transport from the cytoplasm to the chloroplast, which is required for photosynthesis [73]. PHO1 transporters translocate inorganic phosphate from the root to the shoot in plants [74, 75]. They are localized in the plasma membrane of the root endodermis and are responsible for loading inorganic phosphate into the xylem for transport to the shoot [74]. PHO1 transporters have also been shown to play a role in the secretion of phosphate-containing compounds into the rhizosphere, which can increase the availability of P in the soil [69]. The regulation of P transporters is critical for

| Protein name | Function | References |
|---|---|---|
| PHT1 Transporters | Facilitate the uptake of inorganic phosphate (Pi) from the soil into root cells | [71, 77] |
| PHO1 Transporter | Transports Pi from root cells to the xylem for long-distance transport to the shoots | [69] |
| SPX Proteins | Regulate Pi homeostasis by inhibiting the activity of PHT1 transporters and promoting PHO1 transporter expression | [78, 79] |
| PHR1 Transcription Factor | Regulates the expression of genes involved in Pi uptake and remobilization in response to Pi deficiency | [80, 81] |
| miR399 | Represses the expression of PHO2, a negative regulator of Pi uptake, under Pi-deficient conditions | [82, 83] |
| IPS1 | Encodes a non-coding RNA that regulates Pi homeostasis by repressing the expression of PHO2 and promoting Pi uptake | [82, 84] |
| Purple Acid Phosphatases (PAPs) | Involved in the hydrolysis of organic phosphates in the soil, releasing Pi for plant uptake | [7, 85] |
| NRT1.1 Transporter | Involved in the uptake of nitrate and can also transport Pi under Pi-deficient conditions | [79, 86] |
| NLA E3 Ubiquitin Ligase | Regulates the trafficking and degradation of PHT1 transporters in response to Pi availability | [87, 88] |
| RbohD NADPH Oxidase | Involved in Pi sensing and signaling by producing reactive oxygen species (ROS) | [89] |

**Table 2.**
*Major plant transport proteins involved in phosphorus uptake.*

maintaining P-homeostasis in plants [2]. Several factors can influence the expression and activity of P-transporters, including P availability, plant age, and environmental stresses [76]. Under low P-conditions, plants can upregulate the expression of PHT1 transporters, leading to an increase in phosphate uptake [76]. Similarly, under drought stress, the expression of PHT1 transporters can be downregulated, leading to a decrease in phosphate uptake [76]. **Table 2** lists the major transport proteins and highlights their important functions.

### 3.7 Organic phosphorus mineralization

According to Sharpley [90], organic P-compounds comprise a significant proportion of total P in many soils. However, these organic P-compounds are often not readily available to plants. Microorganisms play a crucial role in the mineralization of organic P, which is the process by which organic P-compounds are converted into inorganic P-forms. Several factors influence the rate of organic P-mineralization, including soil pH, moisture content, temperature, and the availability of nutrients such as nitrogen and P. For example, studies have shown that organic P-mineralization rates increase with increasing soil pH [91, 92]. Plants can also play a role in organic P-mineralization by releasing organic acids or other compounds that can stimulate microbial activity and increase the availability of inorganic P. For instance, in one experiment the release of root exudates by maize plants increased the mineralization of organic P in the soil [93].

Overall, organic P-mineralization is a crucial process in P-cycling in soils and is essential to plant nutrition. Understanding the factors influencing this process can help optimize fertilizer management strategies and improve crop productivity.

## 4. Phosphorus and microbial communities in the rhizosphere

The availability of P in the soil can significantly affect the composition and function of microbial communities in the rhizosphere. In P-limited soils, microbial communities can be dominated by species adapted to low P-conditions, such as phosphate-solubilizing bacteria and fungi. These microorganisms can produce organic acids and enzymes that can solubilize P-compounds in soil, making them available for plant uptake. On the other hand, high P levels in soil can harm microbial communities in the rhizosphere. Excessive P can lead to the eutrophication of soil and stimulate the growth of opportunistic microorganisms that are not beneficial for plant growth. Moreover, high levels of P can decrease the diversity of microbial communities in soil, as certain species may become dominant due to their ability to tolerate high P-levels.

In addition to the direct impact of P on microbial communities, the composition of microbial communities in the rhizosphere can also affect P-cycling in soil. For example, mycorrhizal fungi in the rhizosphere can enhance P-uptake and translocation in plants, as these fungi can form symbiotic associations with plant roots and improve nutrient uptake.

The availability of P in the soil can strongly influence microbial community composition in the rhizosphere. For example, high P-levels in soil can decrease the abundance of arbuscular mycorrhizal fungi (AMF) [77]. AMF play a crucial role in

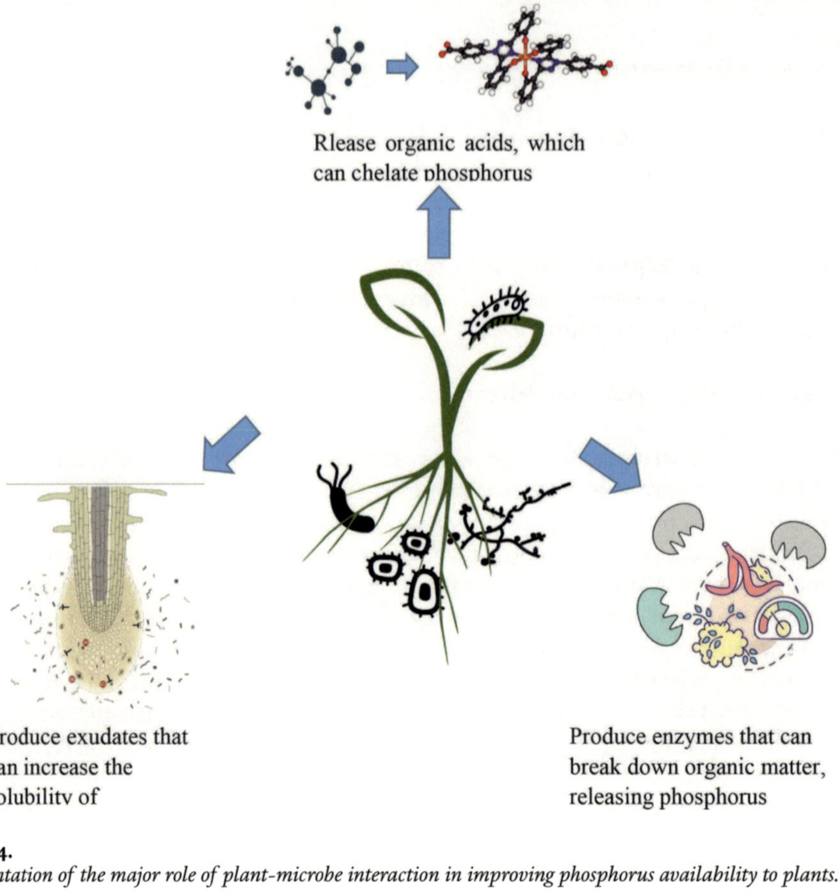


**Figure 4.**
*Representation of the major role of plant-microbe interaction in improving phosphorus availability to plants.*

plant P acquisition, and their reduction can harm plant growth. In contrast, low P levels can increase the abundance of bacteria that can solubilize P [94]. These bacteria can release P from organic compounds and make them available to plants.

Moreover, the form of P in the soil can also affect microbial community composition. Inorganic P, such as phosphate, is soil's most common form of P. However, organic P, such as phytate, is also present in the soil, and its availability is generally low [95]. Some microorganisms, such as phosphate solubilizing bacteria (PSB) and fungi, can convert organic P into inorganic P, making it available to plants. The activity of these microorganisms can be influenced by the plant species and the management practices used in the field [96]. **Figure 4** describes impact of P on microbial communities and plant.

## 5. Plant-microbe interactions and phosphorus-availability in the rhizosphere

One of the most well-known plant-microbe interactions is the mycorrhizal association between plants and fungi (**Figure 5**). Mycorrhizal fungi form a symbiotic relationship with plant roots, where the fungi colonize the roots and extend their hyphae into the soil, increasing the surface area for nutrient uptake. In return, the fungi receive carbohydrates from the plant. Mycorrhizal fungi are particularly effective at accessing and mobilizing P in the soil, often found in low concentrations and insoluble forms. The fungi can release enzymes that break down organic forms of P and make them available for plant uptake [5].

Another type of microbes that can impact P availability in the rhizosphere are PSB. These bacteria convert insoluble forms of P into soluble forms, which plants can take up. PSB can release organic acids, chelating agents, and enzymes that solubilize P, making it available to the plant [97]. In addition to solubilizing P, PSB can enhance root growth and plant biomass by producing phytohormones [98].

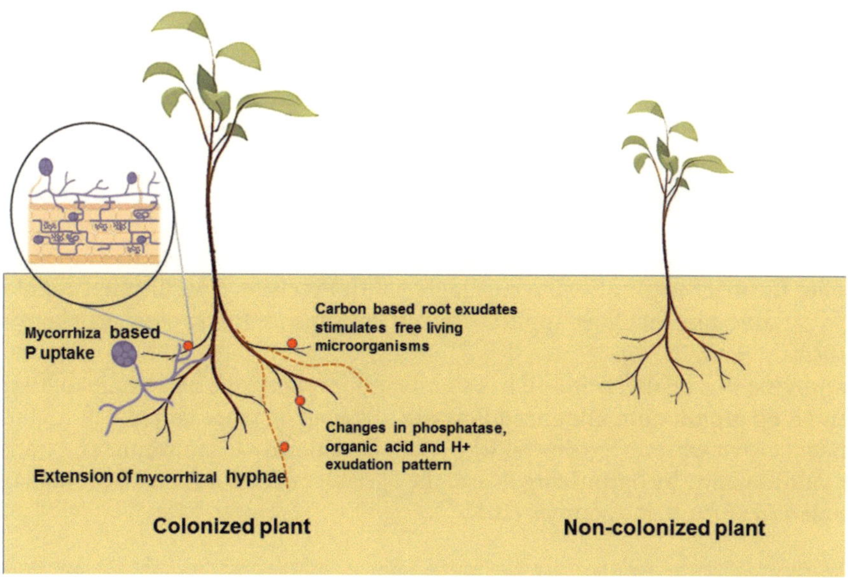


**Figure 5.**
*Schematic representation of the major impacts of mycorrhizal assoctiations on phosphorus uptake on plant.*

Plant-microbe interactions in the rhizosphere can also affect the distribution of P in the soil. For example, some plant species can exude organic compounds that attract specific microbes, influencing the spatial distribution of P in the soil [99]. The presence of microbes in the rhizosphere can also alter the chemical properties of the soil, making it more favorable for plant growth and P uptake [100]. Research has also shown that PSB can interact synergistically with other soil microorganisms, such as mycorrhizal fungi, to further enhance plant growth and nutrient uptake. By promoting a diverse and healthy microbial community in the rhizosphere, PSB can help to create a more resilient and sustainable agricultural system.

## 6. Improved phosphorus-availability in soil: implications for sustainable agriculture and crop production

The PSB are known to solubilize P from organic and inorganic sources and make P more available for plant uptake [6]. When introduced into the rhizosphere, PSB can improve the availability of P in soil and enhance plant growth and development [101]. One potential implication of using PSB in sustainable agriculture is the reduced need for synthetic fertilizers. Synthetic fertilizers are commonly used to supplement soil P levels but can have negative environmental impacts such as eutrophication and groundwater pollution [102]. Using PSB, farmers can reduce their reliance on synthetic fertilizers and promote more sustainable agricultural practices. Another implication of using PSB is the potential for improved crop yields and food security. The availability of P in soil is a limiting factor for crop production. PSB can increase the amount of available P and enhance crop growth and development [103], leading to higher crop yields, improved food security, and increased farmer income.

The P resources are finite, and the increasing global demand for food has pressured this vital nutrient's availability [104]. PSMs are group of microorganisms such as bacteria, fungi, and actinomycetes, that solubilize insoluble phosphate compounds [6]. PSMs as biofertilizers have been found to improve crop yields and P use-efficiency [32]. In addition, incorporating organic materials, such as compost and manure, can enhance soil P availability by increasing microbial activity and soil organic matter content [105]. Organic amendments can also help buffer soil pH, increasing P-availability [106]. Also, breeding plants with improved P-uptake and -utilization capabilities can produce more efficient P-use and higher yields [107]. This approach requires genetic resources and a deeper understanding of the mechanisms underlying plant P-uptake and -utilization [108]. Nevertheless, through precision agriculture and soil testing, farmers can better understand the P-status of their soil and tailor fertilizer applications accordingly [109]. Precision agriculture technologies, such as variable-rate fertilizer applicators, can help ensure P is applied efficiently and effectively.

Improving P-availability in soil has significant implications for sustainable agriculture and crop production. Enhanced P-availability can increase crop yields and reduce dependence on synthetic fertilizers, which can have negative environmental impacts [110]. Additionally, by optimizing P-use, the agricultural sector can better manage the depletion of finite P resources [104].

## Author details

Anurag Yadav[1*] and Kusum Yadav[2]

1 Department of Microbiology, College of Basic Science and Humanities, Sardarkrushinagar Dantiwada Agricultural University, Banaskantha, Gujarat, India

2 Department of Biochemistry, University of Lucknow, Lucknow, Uttar Pradesh, India

*Address all correspondence to: anuragyadav123@gmail.com

## References


[1] Berg G, Smalla K. Plant species and soil type cooperatively shape the structure and function of microbial communities in the rhizosphere. FEMS Microbiology Ecology. 2009;**68**(1):1-13

[2] Raghothama KG. Phosphate Acquisition. Annual Review of Plant Physiology and Plant Molecular Biology. 1999;**50**(1):665-693

[3] Hinsinger P. Bioavailability of soil inorganic P in the rhizosphere as affected by root-induced chemical changes: A review. Plant and Soil. 2001;**237**(2):173-195

[4] Richardson AE et al. Acquisition of Phosphorus and Nitrogen in the Rhizosphere and Plant Growth Promotion by Microorganisms. Plant Soil. 2009;**321**:305-339

[5] Smith S, Read D. Mycorrhizal Symbiosis. Cambridge, UK: Academic Press Ltd; 2008

[6] Rodríguez, H. and R. Fraga, Phosphate solubilizing bacteria and their role in plant growth promotion. Biotechnology Advances, 1999. 17(4-5): p. 319-339.

[7] Richardson AE, Hadobas PA, Hayes JE. Extracellular secretion of *Aspergillus* phytase from Arabidopsis roots enables plants to obtain phosphorus from phytate. The Plant Journal. 2001;**25**(6):641-649

[8] Barea J-M et al. Microbial co-operation in the rhizosphere. Journal of Experimental Botany. 2005;**56**(417):1761-1778

[9] Tilman D et al. Agricultural sustainability and intensive production practices. Nature. 2002;**418**(6898):671-677

[10] Vance CP, Uhde-Stone C, Allan DL. Phosphorus acquisition and use: Critical adaptations by plants for securing a nonrenewable resource. New Phytologist. 2003;**157**(3):423-447

[11] Schachtman DP, Reid RJ, Ayling SM. Phosphorus uptake by plants: From soil to cell. Plant Physiology. 1998;**116**(2):447-453

[12] Richardson A. Soil microorganisms and phosphorus availability. Soil Biota: Management in Sustainable Farming Systems. 1994;**1994**:50-62

[13] Smith SE, Smith FA, Jakobsen I. Mycorrhizal fungi can dominate phosphate supply to plants irrespective of growth responses. Plant Physiology. 2003;**133**(1):16-20

[14] Shen J et al. Phosphorus dynamics: From soil to plant. Plant Physiology. 2011;**156**(3):997-1005

[15] Neumann G, Romheld V. The release of root exudates as affected by the plant's physiological status. In: The Rhizosphere. USA, Switzerland: CRC Press; 2000. pp. 57-110

[16] Barea J-M et al. Mycorrhizal symbioses. The Ecophysiology of Plant-Phosphorus Interactions. 2008;**2008**:143-163

[17] Rao IM, Pessarakli M. The role of phosphorus in photosynthesis. Handbook of Photosynthesis. 1996;**1996**:173-194

[18] Marschner P. Mineral Nutrition of Higher Plants. Amsterdam, Netherlands: Academic; 2012

[19] Michell RH. Inositol derivatives: Evolution and functions. Nature

Reviews Molecular Cell Biology. 2008;**9**(2):151-161

[20] Tariq A et al. Regulation of metabolites by nutrients in plants. Plant Ionomics: Sensing, Signaling, and Regulation. 2023;**2023**:1-18

[21] Malhotra H et al. Phosphorus nutrition: Plant growth in response to deficiency and excess. Plant Nutrients and Abiotic Stress Tolerance. 2018:171-190

[22] Lynch JP. Root phenes for enhanced soil exploration and phosphorus acquisition: Tools for future crops. Plant Physiology. 2011;**156**(3):1041-1049

[23] Ha S, Tran L-S. Understanding plant responses to phosphorus starvation for improvement of plant tolerance to phosphorus deficiency by biotechnological approaches. Critical Reviews in Biotechnology. 2014;**34**(1):16-30

[24] Gaind S, Gaur AC. Thermotolerant phosphate solubilizing microorganisms and their interaction with mung bean. Plant and Soil. 1991;**133**(1):141-149

[25] Xiao C et al. Evaluation for rock phosphate solubilization in fermentation and soil–plant system using a stress-tolerant phosphate-solubilizing *Aspergillus niger* WHAK1. Applied Biochemistry and Biotechnology. 2013;**169**(1):123-133

[26] Asuming-Brempong S, Anipa B. The use of rock phosphate and phosphate solubilising fungi (*Aspergillus niger*) to improve the growth and the yield of upland rice on typic kandiudalf. West African. Journal of Applied Ecology. 2014;**22**(1):27-39

[27] Tahir M et al. Isolation and identification of phosphate solubilizer "Azospirillum, Bacillus" and "Enterobacter" strains by 16SrRNA sequence analysis and their effect on growth of wheat ('*Triticum aestivum*' L.). Australian Journal of Crop Science. 2013;**7**(9):1284-1292

[28] Rajaee S, Alikhani H, Raiesi F. Effect of plant growth promoting potentials of Azotobacter chroococcum native strains on growth, yield and uptake of nutrients in wheat. Isfahan University of Technology-Journal of Crop Production and Processing. 2007;**11**(41):285-297

[29] Kundu BS, Gaur AC. Establishment of nitrogen-fixing and phosphate-solubilising bacteria in rhizosphere and their effect on yield and nutrient uptake of wheat crop. Plant and Soil. 1980;**57**(2):223-230

[30] Wang K et al. Improved grain yield and lowered arsenic accumulation in rice plants by inoculation with arsenite-oxidizing Achromobacter xylosoxidans GD03. Ecotoxicology and Environmental Safety. 2020;**206**:111229

[31] Khan MR, Khan SM. Effects of root-dip treatment with certain phosphate solubilizing microorganisms on the fusarial wilt of tomato. Bioresource Technology. 2002;**85**(2):213-215

[32] Sharma SB et al. Phosphate solubilizing microbes: Sustainable approach for managing phosphorus deficiency in agricultural soils. Springerplus. 2013;**2**:1-14

[33] El-Komy HM. Coimmobilization of Azospirillum lipoferum and Bacillus megaterium for successful phosphorus and nitrogen nutrition of wheat plants. Food Technology and Biotechnology. 2005;**43**(1):19-27

[34] Shahid M, Khan MS. Glyphosate induced toxicity to chickpea plants and stress alleviation by herbicide tolerant phosphate solubilizing *Burkholderia*

*cepacia* PSBB1 carrying multifarious plant growth promoting activities. 3 Biotech. 2018;**8**(2):131

[35] Lee K-E et al. Isolation and characterization of the high silicate and phosphate solubilizing novel strain Enterobacter ludwigii GAK2 that promotes growth in rice plants. Agronomy. 2019;**9**(3):144

[36] Nakmee PS, Techapinyawat S, Ngamprasit S. Comparative potentials of native arbuscular mycorrhizal fungi to improve nutrient uptake and biomass of Sorghum bicolor Linn. Agriculture and Natural Resources. 2016;**50**(3):173-178

[37] Karamanos R, Flore N, Harapiak J. Re-visiting use of *Penicillium bilaii* with phosphorus fertilization of hard red spring wheat. Canadian Journal of Plant Science. 2010;**90**(3):265-277

[38] Viruel E et al. Inoculation of maize with phosphate solubilizing bacteria: Effect on plant growth and yield. Journal of Soil Science and Plant Nutrition. 2014;**14**(4):819-831

[39] Gond S et al. Induction of salt tolerance and up-regulation of aquaporin genes in tropical corn by rhizobacterium Pantoea agglomerans. Letters in Applied Microbiology. 2015;**60**(4):392-399

[40] Schoebitz M, Ceballos C, Ciamp L. Effect of immobilized phosphate solubilizing bacteria on wheat growth and phosphate uptake. Journal of Soil Science and Plant Nutrition. 2013;**13**(1):1-10

[41] Gamalero E et al. Morphogenetic modifications induced by *Pseudomonas* fluorescens A6RI and Glomus mosseae BEG12 in the root system of tomato differ according to plant growth conditions. New Phytologist. 2002;**155**(2):293-300

[42] Pereira NCM et al. Corn yield and phosphorus use efficiency response to phosphorus rates associated with plant growth promoting bacteria. Frontiers in Environmental Science. 2020;**8**:40

[43] Hameeda B et al. Growth promotion of maize by phosphate-solubilizing bacteria isolated from composts and macrofauna. Microbiological Research. 2008;**163**(2):234-242

[44] Singh RP, Jha PN. The multifarious PGPR Serratia marcescens CDP-13 augments induced systemic resistance and enhanced salinity tolerance of wheat (*Triticum aestivum* L.). PLoS One. 2016;**11**(6):e0155026

[45] Suárez-Moreno ZR et al. Plant-growth promotion and biocontrol properties of three Streptomyces spp. isolates to control bacterial rice pathogens. Frontiers in Microbiology. 2019;**10**:290

[46] Kotasthane A et al. In-vitro antagonism of Trichoderma spp. against Sclerotium rolfsii and Rhizoctonia solani and their response towards growth of cucumber, bottle gourd and bitter gourd. European Journal of Plant Pathology. 2015;**141**(3):523-543

[47] Paul S, Rakshit A. Effect of seed bio-priming with Trichoderma viride strain BHU-2953 for enhancing soil phosphorus solubilization and uptake in soybean (Glycine max). Journal of Soil Science and Plant Nutrition. 2021;**21**:1041-1052

[48] Miller SS et al. Molecular control of acid phosphatase secretion into the rhizosphere of proteoid roots from phosphorus-stressed white lupin. Plant Physiology. 2001;**127**(2):594-606

[49] George T et al. Phosphatase activity and organic acids in the rhizosphere of potential agroforestry species and

maize. Soil Biology and Biochemistry. 2002;**34**(10):1487-1494

[50] López-Bucio J, Cruz-Ramırez A, Herrera-Estrella L. The role of nutrient availability in regulating root architecture. Current Opinion in Plant Biology. 2003;**6**(3):280-287

[51] Lambers H et al. Plant-Microbe-Soil Interactions in the Rhizosphere: An Evolutionary Perspective. Plant Soil. 2009;**321**:83-115

[52] Khan AA et al. Phosphorus solubilizing bacteria: Occurrence, mechanisms and their role in crop production. Journal of Agricultural Biological Science. 2009;**1**(1):48-58

[53] Hocking, P.J., Organic Acids Exuded from Roots in Phosphorus Uptake and Aluminum Tolerance of Plants in Acid Soils. Advances in Agronomy. 2001;**74**:63-97

[54] Jones DL. Organic acids in the rhizosphere–a critical review. Plant and Soil. 1998;**205**:25-44

[55] Zhang C et al. The origin and composition of cucurbit "phloem" exudate. Plant Physiology. 2012;**158**(4):1873-1882

[56] Chen S et al. Root-associated microbiomes of wheat under the combined effect of plant development and nitrogen fertilization. Microbiome. 2019;7(1):1-13

[57] Palomo L, Claassen N, Jones DL. Differential mobilization of P in the maize rhizosphere by citric acid and potassium citrate. Soil Biology and Biochemistry. 2006;**38**(4):683-692

[58] Jia Y et al. Identification of two chickpea multidrug and toxic compound extrusion transporter genes transcriptionally upregulated upon aluminum treatment in root tips. Frontiers in Plant Science. 2022;**2022**:13

[59] Bucher M. Functional biology of plant phosphate uptake at root and mycorrhiza interfaces. New Phytologist. 2007;**173**(1):11-26

[60] Lynch JP. Root phenes that reduce the metabolic costs of soil exploration: Opportunities for 21st century agriculture. Plant, Cell & Environment. 2015;**38**(9):1775-1784

[61] López-Bucio J et al. Phosphate availability alters architecture and causes changes in hormone sensitivity in the Arabidopsis root system. Plant Physiology. 2002;**129**(1):244-256

[62] Miguel MA, Postma JA, Lynch JP. Phene synergism between root hair length and basal root growth angle for phosphorus acquisition. Plant Physiology. 2015;**167**(4):1430-1439

[63] Lambers H et al. Plant nutrient-acquisition strategies change with soil age. Trends in Ecology & Evolution. 2008;**23**(2):95-103

[64] Vance CP. Symbiotic nitrogen fixation and phosphorus acquisition. Plant nutrition in a world of declining renewable resources. Plant Physiology. 2001;**127**(2):390-397

[65] Shi R et al. Senescence-induced iron mobilization in source leaves of barley (*Hordeum vulgare*) plants. New Phytologist. 2012;**195**(2):372-383

[66] Veneklaas EJ et al. Opportunities for improving phosphorus-use efficiency in crop plants. New Phytologist. 2012;**195**(2):306-320

[67] Aerts R, Chapin FS III. The mineral nutrition of wild plants revisited: A

re-evaluation of processes and patterns. Advances in Ecological Research. Academic Press. 1999;**30**:1-67

[68] Wu P et al. Phosphate starvation triggers distinct alterations of genome expression in Arabidopsis roots and leaves. Plant Physiology. 2003;**132**(3):1260-1271

[69] Poirier Y, Bucher M. Phosphate transport and homeostasis in Arabidopsis. The Arabidopsis Book/ American Society of Plant Biologists. 2002;**2002**:1

[70] Baker A et al. Replace, reuse, recycle: Improving the sustainable use of phosphorus by plants. Journal of Experimental Botany. 2015;**66**(12):3523-3540

[71] Nussaume L et al. Phosphate import in plants: Focus on the PHT1 transporters. Frontiers in Plant Science. 2011;**2**:83

[72] Raghothama K, Karthikeyan A. Phosphate acquisition. Plant and Soil. 2005;**274**:37-49

[73] Guo B et al. Functional analysis of the Arabidopsis PHT4 family of intracellular phosphate transporters. New Phytologist. 2008;**177**(4):889-898

[74] Hamburger D et al. Identification and characterization of the Arabidopsis PHO1 gene involved in phosphate loading to the xylem. The Plant Cell. 2002;**14**(4):889-902

[75] Smith FW et al. Phosphate transport in plants. Plant and Soil. 2003;**248**:71-83

[76] López-Arredondo DL et al. Phosphate nutrition: Improving low-phosphate tolerance in crops. Annual Review of Plant Biology. 2014;**65**:95-123

[77] Zhang L et al. Carbon and phosphorus exchange may enable cooperation between an arbuscular mycorrhizal fungus and a phosphate-solubilizing bacterium. New Phytologist. 2016;**210**(3):1022-1032

[78] Lin W-Y, Huang T-K, Chiou T-J. Nitrogen limitation adaptation, a target of microRNA827, mediates degradation of plasma membrane–localized phosphate transporters to maintain phosphate homeostasis in Arabidopsis. The Plant Cell. 2013;**25**(10):4061-4074

[79] Liu J et al. A vacuolar phosphate transporter essential for phosphate homeostasis in Arabidopsis. Proceedings of the National Academy of Sciences. 2015;**112**(47):E6571-E6578

[80] Chiou T-J, Lin S-I. Signaling network in sensing phosphate availability in plants. Annual Review of Plant Biology. 2011;**62**:185-206

[81] Wang Y et al. Phosphate uptake and transport in plants: An elaborate regulatory system. Plant and Cell Physiology. 2021;**62**(4):564-572

[82] Pant BD et al. Identification of nutrient-responsive Arabidopsis and rapeseed microRNAs by comprehensive real-time polymerase chain reaction profiling and small RNA sequencing. Plant Physiology. 2009;**150**(3):1541-1555

[83] Lin S-I et al. Regulatory network of microRNA399 and PHO2 by systemic signaling. Plant Physiology. 2008;**147**(2):732-746

[84] Bari R et al. PHO2, microRNA399, and PHR1 define a phosphate-signaling pathway in plants. Plant Physiology. 2006;**141**(3):988-999

[85] Li D et al. Purple acid phosphatases of Arabidopsis thaliana: Comparative

analysis and differential regulation by phosphate deprivation. Journal of Biological Chemistry. 2002;**277**(31):27772-27781

[86] Huang N-C et al. Cloning and functional characterization of an Arabidopsis nitrate transporter gene that encodes a constitutive component of low-affinity uptake. The Plant Cell. 1999;**11**(8):1381-1392

[87] Wang Z et al. Rice SPX1 and SPX2 inhibit phosphate starvation responses through interacting with PHR2 in a phosphate-dependent manner. Proceedings of the National Academy of Sciences. 2014;**111**(41):14953-14958

[88] Wang H et al. OsSIZ1, a SUMO E3 ligase gene, is involved in the regulation of the responses to phosphate and nitrogen in rice. Plant and Cell Physiology. 2015;**56**(12):2381-2395

[89] Foreman J et al. Reactive oxygen species produced by NADPH oxidase regulate plant cell growth. Nature. 2003;**422**(6930):442-446

[90] Sharpley AN. Soil phosphorus dynamics: Agronomic and environmental impacts. Ecological Engineering. 1995;**5**(2-3):261-279

[91] Harrison A. Labile organic phosphorus mineralization in relationship to soil properties. Soil Biology and Biochemistry. 1982;**14**(4):343-351

[92] Chepkwony C et al. Mineralization of soil organic P induced by drying and rewetting as a source of plant-available P in limed and unlimed samples of an acid soil. Plant and Soil. 2001;**234**:83-90

[93] Wang G et al. Simulated root exudates stimulate the abundance of Saccharimonadales to improve the alkaline phosphatase activity in maize rhizosphere. Applied Soil Ecology. 2022;**170**:104274

[94] Hinsinger P et al. Rhizosphere geometry and heterogeneity arising from root-mediated physical and chemical processes. New Phytologist. 2005;**168**(2):293-303

[95] Bünemann EK et al. Soil organic phosphorus and microbial community composition as affected by 26 years of different management strategies. Biology and Fertility of Soils. 2008;**44**:717-726

[96] Yang J, Kloepper JW, Ryu C-M. Rhizosphere bacteria help plants tolerate abiotic stress. Trends in Plant Science. 2009;**14**(1):1-4

[97] Wei Y et al. Effect of organic acids production and bacterial community on the possible mechanism of phosphorus solubilization during composting with enriched phosphate-solubilizing bacteria inoculation. Bioresource Technology. 2018;**247**:190-199

[98] Ahemad M, Kibret M. Mechanisms and applications of plant growth promoting rhizobacteria: Current perspective. Journal of King saud University-Science. 2014;**26**(1):1-20

[99] Hinsinger P et al. Origins of root-mediated pH changes in the rhizosphere and their responses to environmental constraints: A review. Plant and Soil. 2003;**248**:43-59

[100] Marschner P, Crowley D, Rengel Z. Rhizosphere interactions between microorganisms and plants govern iron and phosphorus acquisition along the root axis–model and research methods. Soil Biology and Biochemistry. 2011;**43**(5):883-894

[101] Glick BR. Plant growth-promoting bacteria: Mechanisms and applications. Scientifica. 2012;**2012**:1-15

[102] Sharpley A et al. Kleinman, P1: CAS: 528: DC% 2BC3sXhsVygt73M: Phosphorus legacy: Overcoming the effects of past management practices to mitigate future water quality impairment. Journal of Environmental Quality. 2013;**42**:1308-1326

[103] Kucey R. Increased phosphorus uptake by wheat and field beans inoculated with a phosphorus-solubilizing Penicillium bilaji strain and with vesicular-arbuscular mycorrhizal fungi. Applied and Environmental Microbiology. 1987;**53**(12):2699-2703

[104] Cordell D, Drangert J-O, White S. The story of phosphorus: Global food security and food for thought. Global Environmental Change. 2009;**19**(2):292-305

[105] Garg S, Bahl G. Phosphorus availability to maize as influenced by organic manures and fertilizer P associated phosphatase activity in soils. Bioresource Technology. 2008;**99**(13):5773-5777

[106] Haynes R. Effects of liming on phosphate availability in acid soils: A critical review. Plant and Soil. 1982;**68**:289-308

[107] Lynch JP. Roots of the second green revolution. Australian Journal of Botany. 2007;**55**(5):493-512

[108] Wissuwa M, Gamat G, Ismail AM. Is root growth under phosphorus deficiency affected by source or sink limitations? Journal of Experimental Botany. 2005;**56**(417):1943-1950

[109] Mallarino AP, Wittry DJ. Efficacy of grid and zone soil sampling approaches for site-specific assessment of phosphorus, potassium, pH, and organic matter. Precision Agriculture. 2004;**5**:131-144

[110] Carpenter SR et al. Nonpoint pollution of surface waters with phosphorus and nitrogen. Ecological Applications. 1998;**8**(3):559-568

Chapter 6

# Managing Soil and Plant Nutrients: Role of Microbial Phosphate Solubilisation

*A.A. Asanka Udaya Aberathna, Dilan Amila Satharasinghe, Anura Prasantha Jayasooriya, H. Rasika Niranjana Jinadasa, Sivabalan Manopriya, B.P. Ajith Jayaweera, C. Annesly Nandana Fernando, W.A. Dammika Vijitha Weerathilake, G.A. Prathapasinghe, Janitha Abeywickrema Liyanage and J.M. Krishanthi Jayarukshi Kumari Premarathne*

## Abstract

Phosphorus (P) is one of the macronutrients required for the optimum growth and development of plants. The deficiency of P can be compensated by adding chemical fertilisers, which are expensive and have a negative impact on the ecosystem. Solubilisation of phosphate by microorganisms is an emerging application for eco-friendly and sustainable agriculture practices. This chapter discuss the importance of P for plants, the main problems related to the over-exploitation of natural reserves of P and chemical fertilisers, the diversity of phosphate-solubilising microorganisms (PSM), the ability of microorganisms to solubilise phosphates and key mechanisms of microbial P solubilisation, the capability of microorganisms to formulate phosphate-related nanoparticles, the potential of PSM to develop as commercial level biofertiliser and contribution of PSM for achieving Sustainable Development Goals (SDGs). This chapter will highlight the input of PSM in sustainable agriculture.

**Keywords:** phosphate, Solubilisation, microorganisms, plant, soil

## 1. Introduction

Phosphorous (P) is an essential macronutrient in plants [1, 2] which represents 0.12% of earth's crust [3, 4]. The sources of P are available in both organic and inorganic forms [2, 5], which are non-renewable and cannot be substituted [4, 6]. Rock phosphate (apatite) is the best P source compared to the other P resources such as soil, clay, plant and animal matter [4], which is commonly used for phosphate fertiliser production also [7]. Phosphorous is reaching to be a plant nutrient that will limit crop production in the next millennium. P is a major growth-limiting mineral, and unlike nitrogen, there is no large atmospheric source that can provide biologically available P. Furthermore, the low bioavailability of phosphate limits the efficacy of P fertiliser. As the option to manage the bioavailability of phosphate for plants, phosphate-solubilising microorganisms (PSM) can be an effective, eco-friendly and sustainable tool. PSM contributes to improving plant growth and yield while reducing the need for chemical fertilisers that can have negative impacts on the environment. Additionally, PSM phosphate solubilisation can help to improve the overall health and fertility of the soil, promoting sustainable agriculture practices [8].

## 2. Importance of P for plants

P is engaged in biological processes that are shared by all living species [1, 2, 9]. P, in particular, is the second most limiting macronutrient for plant growth, accounting for 0.2% of plant dry weight [10, 11] and serving as a critical element in animal bodies [12–14]. As a result, P has become a fundamental and necessary component for human well-being [15].

Functions of P in all living organisms included energy metabolism and transmission [1, 9], cell division [2], protein and nucleic acids (DNA and RNA) synthesis [9, 16, 17], being a key component of enzymes, coenzymes, and phospholipids [18] and cell division [2]. In addition, P is essential for plants for photosynthesis [2], nitrogen fixation [14], improving crop quality [14, 16], development of disease resistance [18], root development [3], development of the stems and stalks [15], flower and seed formation [10], conversion of sugar to starch [10], laying down the primordia of propagative parts during the early stages of plant development [16], and proper stress reduction and maturation of plant [10]. A lack of phosphorus can lead to stunted growth, delayed maturity, and lower yield. Therefore, it is essential to ensure that plants receive adequate phosphorus to maintain their health and productivity.

The world population is expected to exceed 9000 million in 2050 [19] and food production requires to be increased globally by 50% as compared to the demand in 2012 [20]. Usability of phosphate has been severely limited due to its low bioavailability. Furthermore, the wastage of phosphate that is unable to be utilised by plants and animals causes additional issues such as eutrophication [10, 21]. The low bioavailability of phosphate is driving its increasing usage, with peak extraction occurring in 2030 [22]. Therefore, the sustainable utilisation of phosphate is needed to increase its bioavailability to avoid the wastage of phosphate [4]. Using microorganisms to increase the bioavailability of phosphate is the emerging eco-friendly technique for sustainable utilisation of phosphate [10, 23].

## 3. Main problems related to the over-exploitation of natural reserves of P and chemical fertilisers

Global food production needs to be increased due to the rapidly increasing world population [20]. The low bioavailability of phosphate directly affects to not being able to fulfil the required amount of crop production. Due to the low bioavailability, an extra amount of phosphate has to be utilised as fertilisers to provide the optimum requirement of P of plants. Therefore, it is directed to high utilisation, and the peak extraction of phosphate will take place in 2030 [22]. Besides the depletion of natural P resources, over-exploitation of natural reserves of P and excessive use of chemical P fertilisers lead to several environmental and agricultural problems such as soil degradation, groundwater contamination, eutrophication and climate change [10, 21]. Hence, it is highly required for sustainable utilisation of phosphate to minimise wastage [4]. State-of-the-art technique to use microorganisms to solubilise the insoluble forms of phosphates and increase their bioavailability is an effective option [10, 23].

## 4. Diversity of phosphate-solubilising microorganisms (PSM)

PSM are diverse and can be found in various environments such as water, and plant tissues while the soil is the key habitat [10, 12]. The population of PSMs in soil ranges between $10^4$ and $10^6$ $g^{-1}$ of soil, and they are accumulated at the rhizosphere of plants, and those organisms are highly metabolically active [10, 24]. In the microbial profile of soil, phosphate-solubilising bacteria (PSB) were responsible for 1–50%, whereas phosphate-solubilising fungi (PSF) represent 0.1–0.5%. The population density of PSB ranged between $8 \times 10^5$ and $5.33 \times 10^9$ in the different rhizospheres of vegetable fields, with PSB accounting for 3.98% of the total population of bacteria [25]. Most of PSBs are coccus, bacillus or spirillum in shape while the bacillus is the most abundant and spirillum are rare. However, there is a high diversity of PSM in the soil. Among these microbial species, *Bacillus* and *Pseudomonas* are the foremost bacterial genera [4, 12, 24].

The number of PSM is not sufficiently high to compete with other microbial species in the rhizosphere. The profile and population of these PSM vary between ecosystems due to the influence of complex biological factors [25]. Physical-chemical factors of soil such as the soil nutrient status, moisture content, organic matter, soil pH, and soil enzyme activities affect these variations [4, 12, 24]. The diversity of PSM (**Table 1**)

| Microorganisms | Country | References |
|---|---|---|
| *Achromobacter xylosoxidans* | China | [26] |
| *Acinetobacter calcoaceticus* | Iran | [27] |
| *Advenella mimigardefordensis* | Spain | [28] |
| *Arthrobacter luteolus* | Iran | [27] |
| *Aspergillus awamori* | Indonesia | [29] |
| *Aspergillus niger* | India, China | [30, 31] |
| *Aspergillus terreus* | Indonesia | [29] |
| *Bacillus amyloliquefaciens* | Italy | [32] |
| *Bacillus aryabhattai* IA20 | Pakistan | [33] |
| *Bacillus cereus* | Spain, Iran | [27, 28] |
| *Bacillus cereus* MZUTZ01 | India | [34] |
| *Bacillus firmus* | Pakistan | [35] |
| *Bacillus licheniformis* | Pakistan | [35] |
| *Bacillus megaterium* | China, Brazil, Spain, Thailand | [28, 36–39] |
| *Bacillus mojavensis* | Thailand | [36] |
| *Bacillus pumilus* | Mexico | [37] |
| *Bacillus safensis* | Pakistan | [35] |
| *Bacillus safensis* IALR1035 | USA | [38] |

| Microorganisms | Country | References |
|---|---|---|
| *Bacillus siamensis* | Mexico | [37] |
| *Bacillus subtilis* | India, Indonesia, Brazil | [29, 34, 39] |
| *Bacillus subtilis* IA6 | Pakistan | [33] |
| *Bacillus subtilis* IALR1033 | USA | [38] |
| *Bacillus thuringensis* MZUTZ13 | India | [34] |
| *Burkholderia cenocepacia* | Indonesia | [29] |
| *Burkholderia cepacia* | Indonesia | [29] |
| *Burkholderia cepacia* ISOP5 | China | [40] |
| *Burkholderia fungorum* | Spain | [28] |
| *Burkholderia gladioli* | India | [34] |
| *Burkholderia seminalis* | Indonesia | [29] |
| *Burkholderia vietnamiensis* | Iran | [27] |
| *Cellulosimicrobium cellulans* | China | [26] |
| *Enterobacter bugandensis* | Morocco | [41] |
| *Enterobacter cloacae* C8 | China | [42] |
| *Enterobacter hormaechei* (LMG 27195) | China | [43] |
| *Funneliformis mosseae* | China | [44] |
| *Funneliformis mosseae* BEG234 | Italy | [32] |
| *Geobacillus stearothermophillus* MZUTZ08 | India | [34] |
| *Klebsiella variicola* | Brazil | [45] |
| *Nocardiopsis alba* | Morocco | [46, 47] |
| *Novosphingobium barchaimii* (LL02) | China | [43] |
| *Novosphingobium resinovorum* | China | [42] |
| *Ochrobactrum haematophilum* | China | [26] |
| *Ochrobactrum pseudogrignonense* | Brazil | [45] |
| *Ochrobactrum pseudogrignonense* (CCUG30717) | China | [43] |
| *Paenibacillus polymyxa* IA7 | Pakistan | [33] |
| *Pantoea agglomerans* | Tunisia, Morocco | [41, 48] |
| *Pantoea agglomerans* IALR1325 | USA | [38] |
| *Pantoea agglomerans* pv. P5 | Iran | [49] |
| *Pantoea ananatis* | Brazil | [45] |
| Pantoea roadsii (LMG26273) | China | [43] |
| *Pantoea stewartii* subsp. *Indologenes* | Morocco | [41] |
| *Pantoea vagans* IALR611 | USA | [38] |
| *Paraburkholderia caffeinilytica* (CF1) | China | [43] |
| *Penicillium oxalicum* | Pakistan | [35] |
| *Pseudomonas agglomerans* | Tunisia | [50] |
| *Pseudomonas azotoformans* | India | [51] |

| Microorganisms | Country | References |
|---|---|---|
| *Pseudomonas brassicacearum* supsp. *Neoaurantiaca* | Morocco | [41] |
| *Pseudomonas cepaceae* | Egypt | [52] |
| *Pseudomonas donghuensis* (HYS) | China | [43] |
| *Pseudomonas fluorescens* | Tunisia | [50] |
| *Pseudomonas grimontii* (CFML97 514) | China | [43] |
| *Pseudomonas lactis* | Morocco | [41] |
| *Pseudomonas libanensis* | Mexico | [37] |
| *Pseudomonas mallei* | Egypt | [52] |
| *Pseudomonas palleroniana* | India | [51] |
| *Pseudomonas plecoglossicida* C10 | China | [42] |
| *Pseudomonas proteolytica* | India | [51] |
| *Pseudomonas psychrotolerans* IALR632 | USA | [38] |
| *Pseudomonas putida* pv. P13 | Iran | [49] |
| *Ralstonia pickettii* C9 | China | [42] |
| *Rhizophagus irregularis* BEG72 | Italy | [32] |
| *Rhodopseudomonas palustris* ISP-1 | China | [40] |
| *Serratia rubidaea* | Morocco | [41] |
| *Staphylococcus pastueri* MZUTZ02 | India | [34] |
| *Stenotrophomonas maltophilia* | Tunisia, China | [26, 50] |
| *Streptomyces alboviridis* | Morocco | [47] |
| *Streptomyces fulvissimus* | Morocco | [46] |
| *Streptomyces griseorubens* | Morocco | [46, 47] |
| *Streptomyces microflavus* | Morocco | [46] |
| *Streptomyces pratensis* | Morocco | [46] |
| *Streptomyces youssoufiensis* | Morocco | [46] |
| *Talaromyces minioluteus* | Iran | [27] |
| *Talaromyces pinophilus* | Iran, Indonesia | [27, 29] |
| *Talaromyces stipitatus* | Iran | [27] |
| *Trichoderma asperellum* LZ1 | China | [53] |

**Table 1.**
*Major phosphate solubilizing microorganisms.*

is important for maintaining healthy ecosystems, as they play a critical role in the phosphorus cycle by making this important nutrient more available to plants [25].

## 5. Ability of microorganisms to solubilise phosphates and key mechanisms of microbial P Solubilisation

Phosphorous is available in the soil as a number of organic and inorganic compounds due to its high reactivity, and those are unavailable for plants [54]. PSMs solubilise these insoluble forms of phosphates into soluble by secreting complex compounds such as organic acid anions, protons, exopolysaccharides, siderophores, hydroxyl ions and extracellular enzymes [55]. The phosphate-solubilising mechanism of PSM has two major aspects as inorganic phosphate solubilisation and organic phosphate solubilisation based on the substrate of P [3].

### 5.1 Inorganic phosphate solubilisation by PSM

Microbial mobilisation of inorganic phosphates is involved with the production and secretion of organic acids, inorganic acids, siderophore and exopolysaccharide and proton extrusion by the PSM [3].

#### *5.1.1 Organic acid production*

The key mechanism of phosphate solubilisation is the secretion of the organic acids that result from the carbon metabolism of PSM, which is closely related to the concentration of soluble phosphate [3, 4, 24]. Low-molecular-weight organic acids are synthetised during glucose oxidation through direct periplasmic oxidation and intracellular phosphorylation. The synthesis of organic acids from glucose by the cells of PSMs in phosphate deficient conditions is higher than in phosphate sufficient conditions, which correlates with the effect of soluble phosphate on organic acid production [56].

The release of these organic acids into the environment is accompanied by a decrease in pH and chelate of the cations (such as $Al^{3+}$, $Fe^{3+}$, and $Ca^{2+}$) bound to phosphate ions to release the phosphate [3, 8]. These organic acids compete with the phosphate binding sites of the medium and allow phosphates to be available in free [15] as $HPO_4^{-2}$ and $HPO_4^{-3}$ [10]. However, there is no correlation between pH and the amount of solubilised phosphates [56]. Gluconic acid is the frequent secretion among all organic acids released by PSM (**Table 2**) [3, 10, 54]. PSF may be even more important than PSB since they typically produce and excrete more acids [56].

#### *5.1.2 Inorganic acid production*

Inorganic acid-producing bacteria also involve with phosphate solubilisation through acidification. Bacteria, engaged in nitrification and sulphur-oxidation, which have the ability to produce inorganic acids, are involved with phosphate solubilisation by secreting nitric, carbonic, sulphuric [3] and hydrochloric acids [4, 24]. Nitrifying bacteria such as *Nitrosovibrio*, *Nitrosomonas*, *Nitrobacter*, *Nitrosospira* frequently secrete nitric acid [61] and sulphur oxidising bacteria such as *Thiobacillus thiooxidans* [62] produce sulphuric acid directly involve to solubilisation of phosphate which is required to be further studied. Acidification of the media by secreting inorganic acids and $H^+$ substitution reactions release the phosphates by converting insoluble phosphate to its soluble form [4, 24]. However, the efficacy of phosphate solubilisation by inorganic acids is lower than the efficiency of organic acids [3, 8].

#### *5.1.3 Proton extrusion*

Another alternative mechanism to solubilise phosphate is extrusion of proton, which minimise the requirement to synthesis acids [3]. Excretion of $H^+$ through $H_2CO_3$ production, $NH_4^+$ assimilation and liberation of organic acid anions facilitate the solubilisation of phosphates by acidifying the media [3, 8, 10]. The release of $H^+$ to the extracellular surface of PSM through the exchange of cation or the ATPase activity

| Organic acid | PSM | | References |
|---|---|---|---|
| Acetic acid | *Bacillus amyloliquefaciens*<br>*Bacillus atrophaeus*<br>*Bacillus licheniformis*<br>*Chryseomonas Luteola*<br>*Enterobacter aerogenes*<br>*Enterobacter asburiae* | *Enterobacter taylorae*<br>*Kluyvera cryocrescens*<br>*Penibacillus macerans*<br>*Pseudomonas aerogenes*<br>*Vibrio proteolyticus*<br>*Xanthobacter agilis* | [57] |
| Caproic acid | *Actinomadura oligospora*<br>*Bacillus subtilis var.2* | *Citrobacter* sp. | [57] |
| Citric acid | *Arrhrobacter*<br>*Aspergillus flavus*<br>*Aspergillus niger FS 1*<br>*Enterobacter*<br>*Eupenicillium ludwigii FS 27* | *Penicillium canescens*<br>*Penicillium canescens*<br>*FS 23*<br>*Penicillium islandicum*<br>*FS 30*<br>*Penicillium rugulosum*<br>*Penicillium trivialis* | [12] |
| | *Arthrobacter* sp.<br>*Aspergillus flavus*<br>*Aspergillus foetidus*<br>*Aspergillus japonicas*<br>*Aspergillus niger*<br>*Bacillus firmus B-7650* | *Bascillus* sp.<br>*Chaetomium nigricolor*<br>*Enterobacter*<br>*agglomerans*<br>*Penicillium fluorescens*<br>*Penicillium canescens*<br>*Penicillium rugulosum* | [57] |
| | *Aspergillus*<br>*Azospirillum* sp.<br>*Bacillus* sp. | *Penicillium* sp.<br>*Proteus* sp.<br>*Pseudomonas* | [10, 58–60] |
| Formic Acid | *Pseudomonas trivialis* | | [12] |
| | *Actinomadura oligospora*<br>*Bacillus pumilus var.2*<br>*Bacillus subtilis var.2* | *Citrobacter* sp.<br>*Pseudomonas trivialis* | [57] |
| Fumaric acid | *Azospirillum* sp.<br>*Bacillus* sp. | *Proteus* sp.<br>*Pseudomonas* | [10, 59, 60] |
| Gluconic acid | *Arrhrobacter*<br>*Aspergillus flavus*<br>*Aspergillus niger FS 1*<br>*Enterobacter*<br>*Enterobacter intermedium*<br>*Enterobacter sps Fs 11* | *Eupenicillium ludwigii*<br>*FS 27*<br>*Penicillium canescens*<br>*FS23*<br>*Penicillium islandicum*<br>*Penicillium rugulosum*<br>*Pseudomonas*<br>*fluorescens* | [12] |
| | *Actinomadura oligospora*<br>*Aspergillus flavus*<br>*Aspergillus foetidus*<br>*Aspergillus japonicas*<br>*Aspergillus niger*<br>*Bacillus pumilus var.2* | *Bacillus subtilis var.2*<br>*Citrobacter* sp.<br>*Penicillium canescen*<br>*Penicillium radicum*<br>*Penicillium rugulosum*<br>*Pseudomonas*<br>*fluorescens* | [57] |
| | *Aspergillus*<br>*Azospirillum* sp.<br>*Bacillus* sp.<br>*Erwinia herbicola* | *Penicillium* sp.<br>*Proteus* sp.<br>*Pseudomonas* | [10, 58–60] |
| Glycolic acid | *Aspergillus niger* | *Penicillium* sp. | [12] |
| | *Aspergillus* sp. | *Penicillium* sp. | [10, 58] |

| Organic acid | PSM | | References |
|---|---|---|---|
| Heptonic acid | *Actinomadura oligospora*<br>*Bacillus pumilus var.2* | *Bacillus subtilis var.2*<br>*Citrobacter* sp. | [57] |
| Indole acetic acid | *Psedomonas nitroreducens* | | [12] |
| Isobutyric acid | *Bacillus amyloliquefaciens*<br>*Bacillus atrophaeus*<br>*Bacillus licheniformis*<br>*Chryseomonas luteola*<br>*Enterobacter aerogenes*<br>*Enterobacter asburiae* | *Enterobacter taylorae*<br>*Kluyvera cryocrescens*<br>*Penibacillus macerans*<br>*Pseudomonas aerogenes*<br>*Vibrio proteolyticus*<br>*Xanthobacter agilis* | [57] |
| Isocaproic acid | *Actinomadura oligospora*<br>*Bacillus pumilus var.2* | *Bacillus subtilis var.2*<br>*Citrobacter* sp. | [57] |
| Isovaleric acid | *Actinomadura oligospora*<br>*Bacillus atrophaeus*<br>*Bacillus licheniformis*<br>*Bacillus pumilus var.2*<br>*Bacillus subtilis var.2*<br>*Bacillus amyloliquefaciens*<br>*Chryseomonas luteola*<br>*Citrobacter* sp. | *Enterobacter asburiae*<br>*Enterobacter taylorae*<br>*Enterobacter aerogenes*<br>*Kluyvera cryocrescens*<br>*Penibacillus macerans*<br>*Pseudomonas aerogenes*<br>*Vibrio proteolyticus*<br>*Xanthobacter agilis* | [57] |
| Itaconic acid | *Bacillus atrophaeus*<br>*Bacillus licheniformis*<br>*Bacillus amyloliquefaciens*<br>*Chryseomonas luteola*<br>*Enterobacter asburiae*<br>*Enterobacter taylorae* | *Enterobacter aerogenes*<br>*Kluyvera cryocrescens*<br>*Penibacillus macerans*<br>*Pseudomonas aerogenes*<br>*Vibrio proteolyticus*<br>*Xanthobacter agilis* | [57] |
| Lactic acid | *Aspergillus niger*<br>*Penicillium* sp. | *Pseudomonas trivialis* | [12] |
| | *Arthrobacter* sp.<br>*Aspergillus niger*<br>*Bacillus atrophaeus*<br>*Bacillus licheniformis*<br>*Bacillus amyloliquefaciens*<br>*Bacillus firmus B-7650*<br>*Bacillus megaterium*<br>*Chryseomonas luteola*<br>*Enterobacter asburiae*<br>*Enterobacter taylorae* | *Enterobacter aerogenes*<br>*Escherichia freundii*<br>*Kluyvera cryocrescens*<br>*Pseudomonas trivialis*<br>*Penibacillus macerans*<br>*Penicillium* sp.<br>*Pseudomonas aerogenes*<br>*Bacillus subtilus*<br>*Vibrio proteolyticus*<br>*Xanthobacter agilis* | [57] |
| Malic acid | *Arrhrobacter*<br>*Enterobacter sps Fs 11* | *Pseudomonas*<br>*fluorescens* | [12] |
| | *Bacillus megaterium*<br>*Bacillus subtilus* | *Pseudomonas*<br>*fluorescens* | [57] |
| | *Aspergillus*<br>*Bacillus* sp.<br>*Penicillium* sp. | | [10, 58, 59] |
| Malonic acid | *Actinomadura oligospora*<br>*Bacillus pumilus var.2* | *Bacillus subtilis var.2*<br>*Citrobacter* sp. | [57] |
| Oxalacetic acid | *Actinomadura oligospora*<br>*Bacillus pumilus var.2* | *Bacillus subtilis var.2*<br>*Citrobacter* sp. | [57] |

| Organic acid | PSM | | References |
|---|---|---|---|
| Oxalic acid | *A. niger FS 1*<br>*Aspergillus flavus*<br>*Eupenicillium ludwigii FS 27* | *Penicillium canescens FS23*<br>*Penicillium canescens,*<br>*Penicillium islandicum* | [12] |
| | *Actinomadura oligospora*<br>*Aspergillus flavus*<br>*Aspergillus foetidus*<br>*Aspergillus japonicas*<br>*Aspergillus niger*<br>*Bacillus pumilus var.2* | *Bacillus subtilis var.2*<br>*Chaetomium nigricolor*<br>*Citrobacter* sp.<br>*Enterobacter agglomerans*<br>*Penicillium canescens* | [57] |
| | *Aspergillus* sp. | *Penicillium* sp. | [10, 58] |
| Propionic acid | *Actinomadura oligospora*<br>*Bacillus pumilus var.2* | *Bacillus subtilis var.2*<br>*Citrobacter* sp. | [57] |
| Succinic acid | *Aspergillus flavus*<br>*Aspergillus niger* | *Penicillium canescens* | [12] |
| | *Actinomadura oligospora*<br>*Aspergillus flavus*<br>*Aspergillus foetidus*<br>*Aspergillus japonicas*<br>*Aspergillus niger* | *Bacillus pumilus var.2*<br>*Bacillus subtilis var.2*<br>*Chaetomium nigricolor*<br>*Citrobacter* sp.<br>*Penicillium canescens* | [57] |
| | *Aspergillus*<br>*Azospirillum* sp.<br>*Bacillus* sp. | *Penicillium* sp.<br>*Proteus* sp.<br>*Pseudomonas* | [10, 58–60] |
| Tartaric acid | *Arrhrobacter*<br>*Enterobacter* | *Pseudomonas trivialis* | [12] |
| | *Aspergillus japonicas*<br>*Aspergillus foetidus* | *Pseudomonas fluorescens* | [57] |
| | *Bacillus* sp. | | [10, 59] |
| Valeric acid | *Actinomadura oligospora*<br>*Bacillus pumilus var.2* | *Bacillus subtilis var.2*<br>*Citrobacter* sp. | [57] |
| 2-Keto gluconic acid | *Enterobacter intermedium* | | [12] |
| | *Aspergillus* sp.<br>*Chaetomiumnigricolor* | *Enterobacter intermedium*<br>*Penicillium* sp. | [57] |
| | *Erwinia herbicola* | *Pseudomonas* | [10, 59, 60] |

**Table 2.**
*Major examples of the production of organic acids by PSM.*

with H+ translocation, lead to the solubilisation of P [56]. *Pseudomonas fluorescens* [63], *Bacillus* sp., *Azospirillum* sp. [64] significantly exhibit this mechanism of $H^+$ extrusion in the process of phosphate solubilisation.

### *5.1.4 Exopolysaccharide production*

Exopolysaccharides are high molecular weight compounds that indirectly affect the solubilisation of P in soil [65]. Microorganisms secrete exopolysaccharides under stress conditions; exhibit the potential to promote phosphate solubilisation [3, 66]. Exopolysaccharides have a strong affinity with the metal ions in the soil, which have formed complexes with phosphates and release those phosphates. There is a positive correlation between the rate of phosphate solubilisation and the concentration of exopolysaccharides [3]. Different exopolysaccharides have different binding affinities with various metals, and there are also different binding strengths between the metals themselves. *Arthrobacter* sp. ArHy-505, *Azotobacter* sp. AzHy-510 and *Enterobacter* sp. EnHy-401 exhibits higher tricalcium phosphate solubilisation capacity by producing exopolysaccharides [65].

#### *5.1.5 Siderophore production*

Siderophore production is a usual ability of microorganisms, while it is used by PSM as another alternative method to solubilise the phosphate [3, 10, 15]. Siderophores are low-molecular-weight secondary metabolites synthesised by PSM that have a strong affinity for inorganic iron and act as metal chelators. Siderophores contain three functional groups hydroxamates, catecholates, and carboxylates, which catalyse the reduction of $Fe^{3+}$ to $Fe^{2+}$. PSM use siderophores to obtain the Fe as the requirement of cellular functions, and during the dissociation of its bond, P is released and available for plants. *Streptomyces* sp. successfully increase the bioavailability of P through the production of sideraphores [65].

### 5.2 Organic phosphate solubilisation mechanisms

#### *5.2.1 Enzyme production*

The content of organic phosphorus in the soil can reach 30–50% of the total amount, which is represented by compounds such as phosphonates, phytic acid, polyphosphonates, sugar phosphates [3, 56], phosphomonoesters, phosphodiesters, phosphotriesters [8], phospholipids and nucleic acids [3, 8]. Generally, these high molecular weight organic compounds are resistant to chemical hydrolysis and need to be converted into soluble ionic phosphate or low-molecular-weight organic phosphates for plant uptake [56].

PSM produce and secrete the enzymes to solubilise the organic phosphates [8]. There are several groups of enzymes are secreted by PSM as [3, 8, 54–56],

1. Non-Specific Acid Phosphatases (NSAPs)

2. Phytases

3. Phosphonatases

4. Carbon–Phosphorus Lyases

NSAPs are a class of enzymes secreted extracellularly or bound to the lipoprotein membranes of PSM. NSAP are generally known as phosphomonoesterases, which dephosphorylate a wide variety of phosphoesters, solubilising around 90% of organic phosphate in soils. The proportion of phosphatases and abundance of P in the soil has a direct relationship, which influences the availability of phosphates to plants.

According to the authors, the activity of NSAPs is more concentrated in the rhizosphere than in other parts of soil and phosphatase activity, P uptake by plants and nodule weight has a positive correlation [56].

Phytase enzymes are phosphatases produced by PSM, which have the capability to hydrolyse phytic acid by acting on the phosphomonoester bonds present in the compound, originating two subgroups, myo-inositol hexaphosphate or phytate. In addition to P, this process makes biologically available other nutrients such as zinc and iron [65].

Phosphatases (phosphonate hydrolases) are enzymes, which promote the breaking of C-P bonds of phosphonates by catalysing this reaction from a group carbonyl electron scavenger that allows heterologous cleavage between nutrients. Phosphonatases act on several substrates, including phosphoenolpyruvate, phosphonoacetate, and phosphoenol-acetaldehyde and make biologically available P for plants [56, 65].

Carbon–phosphorus lyases are a complex of membrane enzymes released by PSM that also allow the making available of P, cleaving the C–P bonds of phosphonates, producing hydrocarbons and inorganic Phosphate. This complex is the leading mechanism for the utilisation of phosphonates by PSM. The enzymes and proteins of C–P lyases are complex and specific to their substrates. *Escherichia coli* and *Enterobacter cloacae* K7 are commonly involved with solubilisation of phosphates by using C–P lyases [65].

## 6. Capability of microorganisms to formulate phosphate-related nanoparticles

Nanotechnology is a state-of-the-art technique of using particles between 1 to 100 nm, which originated as both organic and inorganic forms [67–69]. Nano minerals are currently used as plant fertilisers due to their Nano size, high surface area, and higher solubility in different solvents and penetrative capacity than conventional mineral fertilisers. Reduction of the particle size is directed to an increment of the surface area of particle and the number of particles per unit in fertiliser, which facilitate higher nutrient utilising efficacy. Nanoparticles enter into plants through nano and microscale openings, which are most commonly available in roots and leaves [70, 71]. Generally, root tips, rhizodermis, lateral root junctions and wounding of roots are the entering pathways to nanoparticles. Enhanced bioavailability of nano fertilisers than conventional fertilisers cause to prevent loss of nutrients from leaching, denitrification, volatilisation, and fixation in the soil to confirm the sustainable utilisation of minerals [70].

Microorganisms including bacteria, fungi, actinobacteria and viruses have the ability to synthesise phosphorus-related nanoparticles with well-defined chemical composition, morphology and size either intra-or extracellularly [71–73]. But no in-depth study or work has been done so far [74]. Enzymes [75, 76] such as phytase, phosphatases [75], extracellular polymeric substances (EPS) [76] and proteins [75] are the key potential materials to synthesise phosphorus-related nanoparticles which are produced by microorganisms. In addition, proteins act as capping agents to stabilise the produced nanoparticles [75]. This is a new arising area in nano formulations [74]. Therefore, limited research has been conducted on synthesising phosphorus-related nanoparticles by using microorganisms, as mentioned in **Table 3**.

| Microbe | Type of Nano P | Size/Shape | Application | References |
|---|---|---|---|---|
| *Aspergillus flavus TFR-1* | Phosphorous | 17–64 nm | Plant fertiliser | [77] |
| *Aspergillus fumigatus* TFR-8 | Phosphorous | 37-81 nm | Plant fertiliser | [77] |
| *Aspergillus oryzae TFR-9* | Phosphorous | 23–88 nm | Plant fertiliser | [77] |
| *Aspergillus terreus CZR-1* | Phosphorous | 11–74 nm | Plant fertiliser | [77] |
| *Aspergillus tubingensis TFR-3* | Phosphorous | 5–49 nm | Plant fertiliser | [77] |
| *Bacillus megaterium JCT13* | Phosphorous | 20–91 nm | To maintenance of soil fertility | [73, 75] |
| *Emericella nidulans TFR-14* | Phosphorous | 55–92 nm | Plant fertiliser | [77] |
| *Rhizoctonia bataticola TFR-6* | Phosphorous | 12–36 nm | Plant fertiliser | [77] |
| Yeast | Zinc Phosphate | 10–80 nm in width and 80–200 nm in length, Butterfly like shape | Antirust pigment and electronic luminophore | [78] |

**Table 3.**
*Microbial synthesis of phosphorus-related nanoparticles.*

## 7. Potential of PSM to develop as a commercial level biofertiliser

The distinctive and unique ability of PSMs to solubilise organic and inorganic phosphates, leads to their developed as commercial level bio fertilisers [10, 12]. Biofertilisers are microbiologically active, eco-friendly, low-cost products applied to soil expecting soil health and growth promotion of plants [10, 15]. Biofertilisers can be replaced by 50% instead of chemical fertilisers without any reduction of the yield [15]. These phosphate bio fertilisers are not crop-specific and able to be used for any type of plant, expecting growth promotion, high yield and crop quality through optimum phosphate absorption [10]. *Azotobacter* inoculation and *Bacillus* inoculants have success stories with yield increments of sugarcane and wheat. Application of the combination of *Bacillus megaterium* and *Azotobacter chroococcum* as biofertiliser remark the yield increment by 10–20%. *Bacillus circulans, Bacillus megaterium, Bacillus subtilis*, and *Pseudomonas striata* reach the commercial level in the phosphate biofertiliser industry [10].

Biofilm inoculants and nano-bio inoculants are ultra-modern techniques in biofertiliser formulation technology. Biofilm inoculants are the combination of two microorganisms, while one microorganism colonises over the other microorganism. The second microorganism act as a biotic surface for the first microorganism to develop a metabolically enhanced biofilm rather than a single culture [3]. Even though these relationships affect the high phosphate solubilisation and growth promotion of plants, the abundance in the soil is very low [3, 79]. Therefore, applying these biofilms as biofertilisers through artificial formulation is advantageous in agriculture [3]. *Pleurotus ostreatus*, *Xanthoparmelia mexicana* and *Penicillium* spp. are the most prominent PSMs, which have the potential to be used in biofilms for substantial impact [80].

## 8. Contribution of PSM to achieving sustainable development goals (SDGs)

Collection of 17 interlinked objectives focusing on prosperity for people and the planet called United Nations Sustainable Development Goals (SDGs) able to be enforced with PSMs. Microbial phosphate solubilisation makes way for sustainable development by promoting plant growth, food security, industrial growth, environmental sustainability and water security [81, 82].

Continuous fertilisation to supply the optimum requirement of phosphate for crops due to the low bioavailability is not the appropriate way, and it is a waste of the resource [83]. Responsible consumption of the resource (SDG 12) able to be confirmed by the involvement of PSMs in increasing the bioavailability of phosphate. It reduces the accumulation of excess phosphate with associated other chemicals as fertilisers to soil and contributes to building up sustainable terrestrial ecosystems with reverse land degradation to halt biodiversity loss (SDG 15) [84]. Moreover, the sequestration of phosphate in the soil leads to eutrophication, which causes disruptions of valuable aquatic resources and pollution the water [85]. Therefore, reducing the excess utilisation of phosphate fertilisers by developing PSM as biofertilisers will ensure the sustainable conservation of aquatic resources (SDG 14) and sustainable management of available clean water (SDG 06) [82, 86].

The contribution of PSM to sustainable agriculture through plant growth promotion ensures sufficient food production empowering food security, which promotes public nutrition and ends hunger (SDG 02) [87, 88]. Development of the sustainable agricultural industry (SDG 09) confirm sustainable economic growth (SDG 08) with moving forward to the reduction of poverty (SDG 01) [89]. Society, which reaches economic development by overcoming poverty and hunger, ensures a better living environment promoting a sustainable peaceful society by accessing justice (SDG 16) [90]. The direct and indirect contributions of PSMs to achieve the SDGs affect significantly strengthen the framework of the 2030 agenda of the United Nations [82].

## 9. Conclusion

The application of PSMs as a tool in managing soil and plant nutrients through Solubilising phosphate is an efficient and eco-friendly method. Bacterial and fungal species are majorly involved in the phosphate solubilisation in soil, remarking *Bacillus*, *Pseudomonas*, *Penicillium* and *Aspergillus* species as the most frequent organisms. Even the PSMs are using a number of mechanisms to solubilise the phosphate, organic acid production is the leading mechanism, and gluconic acid is the foremost organic acid involved in the Phosphate immobilisation process. As a cutting-edge technology, PSMs can be used for the formulation of phosphate-related nanoparticles to increase bioavailability. Current arising biofertiliser technology using this potential of PSMs at the commercial level. Biotechnological applications are able to be used to develop the ability of PSMs to obtain maximum output. Furthermore, PSMs are directly and indirectly affecting to achievement SDGs to serve as a "shared blueprint for peace and prosperity for people and the planet, now and into the future".

## Acknowledgements

The paper was supported by the research grant of "Accelerating Higher Education and Development" (AHEAD) under development-oriented research (DOR) of STEM grant number AHEAD/RA3/DOR/WUSL/LAS No.57 funded by the World Bank.

## Declaration of competing interest

The authors declare that they have no known competing financial interests or personal relationships that could have appeared to influence the work reported in this paper.

## Author details

A.A. Asanka Udaya Aberathna[1], Dilan Amila Satharasinghe[2], Anura Prasantha Jayasooriya[2], H. Rasika Niranjana Jinadasa[3], Sivabalan Manopriya[1], B.P. Ajith Jayaweera[1], C. Annesly Nandana Fernando[4], W.A. Dammika Vijitha Weerathilake[1], G.A. Prathapasinghe[1], Janitha Abeywickrema Liyanage[5] and J.M. Krishanthi Jayarukshi Kumari Premarathne[1*]

1 Department of Livestock and Avian Sciences, Faculty of Livestock, Fisheries and Nutrition, Wayamba University of Sri Lanka, Makandura, Sri Lanka

2 Department of Basic Veterinary Sciences, Faculty of Veterinary Medicine and Animal Science, University of Peradeniya, Peradeniya, Sri Lanka

3 Department of Veterinary Pathobiology, Faculty of Veterinary Medicine and Animal Science, University of Peradeniya, Sri Lanka

4 Department of Nano Science Technology, Faculty of Technology, Wayamba University of Sri Lanka, Kuliyapitiya, Sri Lanka

5 Department of Chemistry, Faculty of Science, University of Kelaniya, Dalugama, Sri Lanka

*Address all correspondence to: krissjayaruk@wyb.ac.lk

## References

[1] Aliyat FZ, Maldani M, El GM, Nassiri L, Ibijbijen J. Isolation and characterization of phosphate Solubilising bacteria from phosphate solid sludge of the Moroccan phosphate mines. Open Agric J. 2020;**14**(17):16-24

[2] Chawngthu L, Hnamte R, Lalfakzuala R. Isolation and characterization of Rhizospheric phosphate Solubilising bacteria from wetland isolation and characterization of Rhizospheric phosphate Solubilising bacteria. Geomicrobiology Journal. 2020;**37**:1-10. DOI: 10.1080/01490451.2019.1709108

[3] Prabhu N, Borkar S, Garg S. Phosphate solubilisation by microorganisms: Overview, mechanisms, applications and advances. In: Advances in Biological Science Research. USA: Elsevier Inc.; 2019. pp. 161-176. DOI: 10.1016/B978-0-12-817497-5.00011-2

[4] Samreen S, Kausar S. Phosphorus fertilizer: The original and commercial sources. In: Phosphorus - Recovery and Recycling. Aligarh, India: London, UKIntechOpen; 2019. pp. 81-92

[5] Sharon JA, Hathwaik LT, Glenn GM, Imam SH, Lee CC. Isolation of efficient phosphate Solubilising bacteria capable of enhancing tomato plant growth. Journal of Soil Science and Plant Nutrition. 2016;**16**(2):525-536

[6] Sarikhani MR, Khoshru B, Greiner R. Isolation and identification of temperature tolerant phosphate Solubilising bacteria as a potential microbial fertilizer. World Journal of Microbiology and Biotechnology. 2019;**1**:1-10. DOI: 10.1007/s11274-019-2702-1

[7] Ameen F, Alyahya SA, Alnadhari S, Alasmari H, Wainwright M. Phosphate Solubilising bacteria and fungi in desert soils: Species, limitations and mechanisms. Archives of Agronomy and Soil Science. 2019;**65**(10):1476-3567. DOI: 10.1080/03650340.2019.1566713

[8] Alori ET, Glick BR, Babalola OO. Microbial phosphorus Solubilisation and its potential for use in sustainable agriculture. Frontiers in Microbiology. 2017;**8**(Jun):10-13

[9] Chen Q, Liu S. Identification and characterization of the phosphate-Solubilising bacterium Pantoea sp. S32 in reclamation soil in Shanxi, China. Frontiers in Microbiology. 2019;**10**:1-12

[10] Kalayu G. Phosphate Solubilising microorganisms: Promising approach as Biofertilisers. 2019;**2019**:1-6

[11] Li Y, Zhang J, Zhang J, Xu W, Mou Z. Characteristics of inorganic phosphate-Solubilising bacteria from the sediments of a eutrophic lake. International Journal of Environmental Research and Public Health. 2019;**16**(12):2141

[12] Anand K, Kumari B, Mallick MA. Phosphate Solubilising microbes: An effective and alternative approach as biofertilisers. International Journal of Pharmacy and Pharmaceutical Sciences. 2016;**8**(2):37-40

[13] Pande A, Pandey P, Mehra S, Singh M. Phenotypic and genotypic characterization of phosphate Solubilising bacteria and their efficiency on the growth of maize. Journal, Genetic Engineering & Biotechnology. 2017;**15**(2):379-391. DOI: 10.1016/j.jgeb.2017.06.005

[14] El-hamshary OI, El-hamshary OIM, Bohkari FM, Al-aklouk LA, Noor SO, Najjar AA. Molecular characterization of some phosphate Solubilising microorganisms. Pharmacophore. 2019;**10**(1):37-51

[15] Satyaprakash M, Nikitha T, Reddi EUB, Reddi EUB, Sadhana B. Phosphorous and phosphate Solubilising bacteria and their role in plant nutrition. Int J Curr Microbiol Appl Sci. 2017;**5**(4):2133-2144

[16] Fenta L. Isolation and characterization of phosphate Solubilising bacteria from tomato (Solanum l.) rhizosphere and their effect on growth and phosphorus uptake of the host plant under green house experiment. Lamenew Fenta Under The Guidance of. International Journal of Advanced Research. 2017;**5**(Jan):1-49

[17] Suleman M, Id SY, Rasul M, Yahya M, Atta M, Mirza MS. Phosphate Solubilising bacteria with glucose dehydrogenase gene for phosphorus uptake and beneficial effects on wheat. PLOSONE. 2018;**13**(9):1-28

[18] Yuliantia E, Rakhmawati A. Screening and characterization of phosphate Solubilising bacteria from isolate of thermophilic bacteria screening and characterization of phosphate Solubilising bacteria from isolate of thermophilic bacteria. In: 4th Int Conf Res Implementation, Educ Math Sci (4th ICRIEMS): New York: AIP. 2018. 2017. p. 090015

[19] Azaroual SE, Hazzoumi Z, El N, Abderrahim M, Issam A, Kadmiri M. Role of inorganic phosphate Solubilising bacilli isolated from Moroccan phosphate rock mine and rhizosphere soils in wheat (Triticum aestivum L) phosphorus uptake. Current Microbiology. 2020;**77**:2391-2404. DOI: 10.1007/s00284-020-02046-8

[20] Hii YS, San CY, Lau SW, DMK. Isolation and characterisation of phosphate Solubilising microorganisms from peat. In: Biocatalysis and Agricultural Biotechnology. Vol. 26. Netherlands: Elsevier; 2020. p. 101643. DOI: 10.1016/j. bcab.2020.101643

[21] Paul D, Sinha SN. Biological removal of phosphate using phosphate Solubilising bacterial consortium from synthetic wastewater: A laboratory sale. Environ. Asia. 2015;**8**(1):1-8

[22] Rosemarin A, de Bruijne G, Caldwell I. Peak phosphorus: The next inconvenient truth. Brok. 2009;**200**(15):6-9

[23] Dipak P, Sinha SN. Bacteria showing phosphate Solubilising efficiency in river sediment. Electron J Biosci. 2013;**1**(1):1-5

[24] Khan MS, Zaidi A, Musarrat J, Technology M. In: Khan MS, Musarrat J, editors. Phosphate Solubilising Microorganisms. New York: Springer; 2014. pp. 1-287

[25] Djuuna IAF, Prabawardani S, Massora M. Population distribution of phosphate-Solubilising microorganisms in agricultural soil. Microbes and Environments. 2022;**37**(1):1-8

[26] Ding Y, Yi Z, Fang Y, He S, Li Y, He K, et al. Multi-omics reveal the efficient phosphate-Solubilising mechanism of bacteria on rocky soil. Frontiers in Microbiology. 2021;**12**(December):761972

[27] Beheshti M, Ali H, Ali A, Etesami H, Asadi H, Norouzi M. Rhizosphere Periphytic biofilm and rice rhizosphere phosphate-Solubilising bacteria and fungi: A possible use for activating occluded P in periphytic

biofilms in paddy fields. Rhizosphere. 2021;**19**(June):100395. DOI: 10.1016/j.rhisph.2021.100395

[28] Ibáñez A, Alba Diez-Galán RC, Calvo-Peña C, Barreiro C, Jesús Medina-Turienzo MS-G, JJRC. Using rhizosphere phosphate Solubilising bacteria to improve barley (Hordeum vulgare) plant productivity. Microorganisms. 2021;**9**(8):1619

[29] Sembiring M, Sabrina T. Diversity of phosphate Solubilising bacteria and fungi from andisol soil affected by the eruption of mount Sinabung, North Sumatra. Indonesia. Biodiversitas. 2022;**23**(2):714-720

[30] Alam K, Barman M, Kumar S, Ray P. Release pattern of soil phosphorus as affected by phosphate Solubilising microorganisms in an Ultisol. Pharma Innov J. 2021;**10**(10):823-827

[31] Shao X, Hao W, Konhauser KO, Gao Y, Tang L, Su M, et al. The dissolution of fluorapatite by phosphate-Solubilising fungi: A balance between enhanced phosphorous supply and fluorine toxicity. Environmental Science and Pollution Research. 2021;**28**:69393-69400

[32] Cozzolino V, Monda H, Savy D, Di MV, Vinci G. Cooperation among phosphate-Solubilising bacteria, humic acids and arbuscular mycorrhizal fungi induces soil microbiome shifts and enhances plant nutrient uptake. Chem Biol Technol Agric. 2021;**8**:1-18. DOI: 10.1186/s40538-021-00230-x

[33] Ahmad I, Ahmad M, Hussain A, Jamil M. Integrated use of phosphate-Solubilising Bacillus subtilis strain IA6 and zinc-Solubilising bacillus sp. strain IA16: A promising approach for improving cotton growth. Folia Microbiologia (Praha). 2020;**66**(1):115-125

[34] Khiangte L, Lalfakzuala R. Effects of heavy metals on phosphatase enzyme activity and Indole-3-acetic acid (IAA) production of phosphate Solubilising bacteria effects of heavy metals on phosphatase enzyme activity and Indole-3-acetic. Geomicrobiology Journal. 2021;**38**:494-503. DOI: 10.1080/01490451.2021.1894271

[35] Qarni A, Billah M, Hussain K, Shah SH, Ahmed W, Alam S, et al. Isolation and characterization of phosphate Solubilising microbes from rock phosphate mines and their potential effect for sustainable agriculture. Sustainability. 2021;**13**:2151

[36] Chungopast S, Thongjoo C, Islam AKMM, Yeasmin S. Efficiency of phosphate-Solubilising bacteria to address phosphorus fixation in Takhli soil series : A case of sugarcane cultivation, Thailand efficiency of phosphate-Solubilising bacteria to address phosphorus fixation in Takhli soil series: A case of sugarcane cultivation, Thailand. Plant and Soil. 2021;**18**(March):347-357

[37] Sanchez-gonzalez ME, Mora-herrera ME, Wong-villarreal A, De LN, Sanchez-paz L, Lugo J, et al. Effect of pH and carbon source on phosphate Solubilisation by bacterial strains in Pikovskaya medium. Microorganisms. 2023;**11**:1-11

[38] Mei C, Chretien RL, Amaradasa BS, He Y, Turner A, Lowman S. CharacterizationofphosphateSolubilising bacterial endophytes and plant growth promotion In vitro and in greenhouse. Microorganisms. 2021;**9**(9):1935

[39] Erica A, De SS, Rodrigues M, Barbosa J, Rafael B, Moreira DA, et al. UAV multispectral data: A reliable approach for managing phosphate-Solubilising bacteria in common bean. Agronomy. 2022;**12**(10):2284

[40] Wang Y, Peng S, Hua Q, Qiu C, Wu P. The long-term effects of using phosphate-Solubilising bacteria and photosynthetic bacteria as Biofertilisers on Peanut yield and soil bacteria community. Frontiers in Microbiology. 2021;**12**(July):1-14

[41] Aliyat FZ, Maldani M, El GM, Nassiri L, Ibijbijen J. Phosphate-Solubilising bacteria isolated from phosphate solid sludge and their ability to solubilise three inorganic phosphate forms: Calcium, iron, and aluminum phosphates. Microorganisms. 2022;**10**:1-13

[42] Song C, Wang W, Gan Y, Wang L. Growth promotion ability of phosphate-Solubilising bacteria from the soybean rhizosphere under maize – Soybean intercropping systems. Journal of the Science of Food and Agriculture. 2021;**102**:1430-1442

[43] Chen J, Zhao G, Wei Y, Dong Y, Hou L. Isolation and screening of multifunctional phosphate Solubilising bacteria and its growth - promoting effect on Chinese fir seedlings. Scientific Reports. 2021;**11**:1-13. DOI: 10.1038/s41598-021-88635-4

[44] An X. Optimizing phosphorus application rate and the mixed inoculation of Arbuscular Mycorrhizal fungi and phosphate-Solubilising bacteria can improve the phosphatase activity and organic acid content in alfalfa soil. Sustainability. 2022;**14**(18):11342

[45] Silva UC, Cuadros-orellana S, Silva DRC, Freitas-júnior LF, Fernandes AC, Leite LR, et al. Genomic and phenotypic insights into the potential of rock phosphate Solubilising bacteria to promote millet growth in vivo. Frontiers in Microbiology. 2021;**11**(January):1-17

[46] Soumare A, Boubekri K, Lyamlouli K, Hafidi M, Ouhdouch Y, Kouisni L. From isolation of phosphate Solubilising microbes to their formulation and use as Biofertilisers: Status and needs. Frontiers in Bioengineering and Biotechnology. 2020;7:1-14

[47] Boubekri K, Soumare A, Mardad I, Lyamlouli K, Hafidi M, Ouhdouch Y, et al. The screening of potassium-and phosphate-Solubilising actinobacteria and the assessment of their ability to promote wheat growth parameters. Microorganisms. 2021;**9**(3):1-16

[48] Saadouli I, Mosbah A, Ferjani R, Stathopoulou P, Galiatsatos I. The impact of the inoculation of phosphate-Solubilising bacteria Pantoea agglomerans on phosphorus availability and bacterial community dynamics of a semi-arid soil. Microorganisms. 2021;**9**(8):1661

[49] Aboksari HA, Hashemabadi D, Kaviani B. Effects of an organic substrate on Pelargonium peltatum and improvement of its morphological, biochemical, and flowering parameters by root-inoculated phosphate Solubilising microorganisms. Communications in Soil Science and Plant Analysis. 2021;**52**(15):1772-1789. DOI: 10.1080/00103624.2021.1892735

[50] Amri M, Mateus D, Gatrouni M, Rjeibi MR, Asses N. Co-inoculation with phosphate-Solubilising microorganisms of rock phosphate and Phosphogypsum and their effect on growth promotion and nutrient uptake by ryegrass. Appl Biosci. 2022;**1**:179-197

[51] Adhikari P, Jain R, Sharma A, Pandey A. Plant growth promotion at low temperature by phosphate-Solubilising pseudomonas Spp. isolated from high-altitude Himalayan soil. Microbial Ecology. 2021;**82**(3):677-687

[52] Bamagoos AA, Alharby HF, Belal EE, Khalaf AEA, Abdelfattah MA, Rady MM, et al. Phosphate-Solubilising bacteria as a panacea to alleviate stress effects of high soil $CaCO_3$ content in Phaseolus vulgaris with special reference to P-releasing enzymes. Sustainability. 2021;**13**:1-22

[53] Guo S, Feng B, Xiao C, Wang Q, Chi R. Phosphate-Solubilising microorganisms to enhance phytoremediation of excess phosphorus pollution in phosphate mining wasteland soil. Bioremediation Journal. 2021;**25**:1-15. DOI: 10.1080/10889868.2021.1884528

[54] Liang JL, Liu J, Jia P, Yang T, Zeng Q, Zhang S, et al. Novel phosphate-Solubilising bacteria enhance soil phosphorus cycling following ecological restoration of land degraded by mining. The ISME Journal. 2020;**14**:1600-1613. DOI: 10.1038/s41396-020-0632-4

[55] Ingle KP. Phosphate Solubilising microbes: An overview. Int J Curr Microbiol Appl Sci. 2017;**6**(1):844-852

[56] Timofeeva A, Galyamova M, Sedykh S. Prospects for using phosphate-Solubilising microorganisms as natural fertilizers in agriculture. Plants. 2022;**11**:1-23

[57] Sharma SB, Sayyed RZ, Trivedi MH, Gobi TA. Phosphate Solubilising microbes: Sustainable approach for managing phosphorus deficiency in agricultural soils. Springer Plus. 2013;**2**(1):1-14

[58] Sane SA, Mehta SK. Isolation and evaluation of rock phosphate Solubilising fungi as potential bio-fertilizer. J Fertil Pestic. 2015;**6**(2):1-6

[59] Selvi KBPJJA, JJA P, Vijaya V, Saraswathi K. Analyzing the efficacy of phosphate Solubilising microorganisms by iMedPub journals analyzing the efficacy of phosphate Solubilising microorganisms by enrichment culture techniques abstract. Biochemistry & Molecular Biology Journal. 2017;**3**(1):1-7

[60] Kumar A, Kumar A, Patel H. Role of microbes in phosphorus availability and acquisition by plants. Int J Curr Microbiol Appl Sci. 2018;**7**(5):1344-1347

[61] Sand W, Ahlers B, Bock E. The impact of microorganisms - especially nitric acid producing bacteria - on the deterioration of natural stones. In: Science, Technology, and European Cultural Heritage: Proceedings of the European Symposium, Bologna, Italy, 13-16 June. UK: Butterworth-Heinemann Publishers; 1991. pp. 481-484. DOI: 10.1016/B978-0-7506-0237-2.50072-1

[62] Liu HL, Lan YW, Cheng YC. Optimal production of sulphuric acid by Thiobacillus thiooxidans using response surface methodology. Process Biochemistry. 2004;**39**(12):1953-1961

[63] Park KH, Lee CY, Son HJ. Mechanism of insoluble phosphate Solubilisation by Pseudomonas fluorescens RAF15 isolated from ginseng rhizosphere and its plant growth-promoting activities. Letters in Applied Microbiology. 2009;**49**(2):222-228

[64] Öğüt M, Er F, Neumann G. Increased proton extrusion of wheat roots by inoculation with phosphorus solubilising microorganims. Plant and Soil. 2011;**339**:285-297

[65] da Silva LI, Pereira MC, de Carvalho AMX, Buttrós VH, PasqualM,DóriaJ.Phosphorus-Solubilising microorganisms: A key to sustainable agriculture. Agriculture. 2023;**13**(2):462

[66] Ionescu M, Belkin S. Overproduction of exopolysaccharides by an

Escherichia coli K-12 rpoS mutant in response to osmotic stress. Applied and Environmental Microbiology. 2009;**75**(2):483-492

[67] Gangadoo S, Stanley D, Hughes RJ, Moore RJ, Chapman J. Nanoparticles in feed: Progress and prospects in poultry research. Trends Food Sci Technol [Internet]. 2016;**58**:115-126. DOI: 10.1016/j.tifs.2016.10.013

[68] Adamo G, Campora S, Ghersi G. Chapter 3 – functionalization of nanoparticles in specific targeting and mechanism release. In: Nanostructures for Novel Therapy. Netherlands: Elsevier Inc.; 2017. pp. 57-80. DOI: 10.1016/B978-0-323-46142-9/00003-7

[69] Adegbeye MJ, Elghandour MMMY, Barbabosa-Pliego A, Monroy JC, Mellado M, Ravi Kanth Reddy P, et al. Nanoparticles in equine nutrition: Mechanism of action and application as feed additives. J Equine Vet Sci. 2019;**78**:29-37. DOI: 10.1016/j.jevs.2019.04.001

[70] Qureshi A, Singh DK, Dwived S. Nano-fertilizers: A novel way for enhancing nutrient use efficiency and crop productivity. Int J Curr Microbiol Appl Sci. 2018;**7**(2):3325-3335

[71] OPK SB, Souto EB. Biosynthesis of Nanonutrients for agricultural applications. In: NanoAgroceuticals & NanoPhytoChemicals. Boca Raton: CRC Press; 2019. pp. 15-30

[72] Swain PS, Rajendran D, Rao SBN, Dominic G. Preparation and effects of nano mineral particle feeding in livestock: A review. Vet World. 2015;**8**(7):888-891

[73] Tarafdar JC, Rathore I. Microbial synthesis of nanoparticles for use in agriculture ecosystem. Microbes Plant Stress Manag. 2016;**2016**(June):105-118

[74] Tarafdar JC, Raliya R, Rathore I. Microbial synthesis of phosphorous nanoparticle from tri-calcium phosphate using aspergillus tubingensis TFR-5. Journal of Bionanoscience. 2012;**6**(2):84-89

[75] Rathore I, Sen M, Gharu AD, Tarafdar JC. An efficient Bm strain JCT13 producing nano- phosphorus particles from phytin and Solubilising phosphates. Int J Curr Eng Technol. 2016;**5**(February):3872-3878

[76] Pirzadah TB, Malik B, Maqbool T, Rehman RU. Development of Nano-bioformulations of nutrients for sustainable agriculture. In: Nanotechnol. Life Sci. Cham: Springer; 2019. pp. 381-394

[77] Tarafdar J, Rathore I, Kaur R, Jain A. Biosynthesis of Nanonutrients for agricultural applications. In: NanoAgroceuticals & NanoPhytoChemicals. New York: CRC Press; 2019. pp. 15-30

[78] Yan S, He W, Sun C, Zhang X, Zhao H, Li Z, et al. The biomimetic synthesis of zinc phosphate nanoparticles. Dyes and Pigments. 2009;**80**(2):254-258. DOI: 10.1016/j.dyepig.2008.06.010

[79] Taktek S, St-arnaud M, Piché Y, Fortin JA, Antoun H. Igneous phosphate rock Solubilisation by biofilm-forming mycorrhizobacteria and hyphobacteria associated with Rhizoglomus irregulare DAOM 197198. Mycorrhiza. 2017;**27**:13-22. DOI: 10.1007/s00572-016-0726-z

[80] Babu SV, Triveni S, Reddy RS, Sathyanarayana J. Persistence of PSB-fungi biofilmed Biofertiliser in the soils and its effect on growth and yield of maize. Int J Curr Microbiol Appl Sci. 2017;**6**(12):1812-1821

[81] Timmis K, de Vos WM, Ramos JL, Vlaeminck SE, Prieto A, Danchin A, et al. The contribution of microbial biotechnology to sustainable development goals. Microbial Biotechnology. 2017;**10**(5):984-987

[82] Aberathna AAAU, Satharasinghe DA, Jayasooriya AP, Jinadasa RN, Manopriya S, Jayaweera BPA, et al. Increasing the bioavailability of phosphate by using microorganisms. Int. J Agron. 2022;**2022**:4305501

[83] Chen X, Yan X, Wang M, Cai Y, Weng X, Su D, et al. Long-term excessive phosphorus fertilization alters soil phosphorus fractions in the acidic soil of pomelo orchards. Soil and Tillage Research. 2022;**215**:105214

[84] Thuy PTP, Hoa NM, Dick WA. Reducing phosphorus fertilizer input in high phosphorus soils for sustainable agriculture in the mekong delta, Vietnam. Agriculture. 2020;**10**(3):87

[85] Oliveira M, Machado AV. The role of phosphorus on eutrophication: A historical review and future perspectives. Environ Technol Rev. 2013;**2**(1):117-127

[86] Wimalawansa SA, Wimalawansa SJ. Protection of watersheds, and control and responsible use of Fertiliser to prevent phosphate eutrophication of reservoirs. Int J Res Environ Sci. 2015;**1**(2):1-18

[87] El Bilali H, Callenius C, Strassner C, Probst L. Food and nutrition security and sustainability transitions in food systems. Food and Energy Security. 2019;**8**(2):1-20

[88] Wu JL, Chen Y, Zhang L, Zhang Y, Wang S, Xin Shi LL, et al. Effects of phosphate Solubilising bacteria on the growth, photosynthesis, and nutrient uptake of Camellia oleifera Abel. Forests. 2019;**10**(348):1-10

[89] Musvoto C, Nortje K, De Wet B, Mahumani BK, Nahman A. Imperatives for an agricultural green economy in South Africa. South African Journal of Science. 2015;**111**(1-2):1-8

[90] Santhirasegaram S. Peace and economic growth in developing countries: Pooled data cross – country empirical study. Peace and economic growth in developing countries: Pooled data cross - country. In: International Conference on Applied Economics – ICOAE. Kastoria, Greece: TEI of Western Macedonia Press; 2008;**2008**. pp. 807-814

Chapter 7

# Phosphorus Recovery through Waste Transformation: Implication for an Alternative Fertilizer

*Valentin Nenov, Hyusein Yemendzhiev and Gergana Peeva*

## Abstract

Presently, the recovery and reuse of phosphorus are still far from being a mainstream practice. Yet, the techniques already accepted and applied differ by the origin of the used matter (wastewater, sludge, ash) and are mainly focused on the process of precipitation. One of these techniques is struvite (magnesium ammonium phosphate; MAP; $MgNH_4PO_4{\cdot}6H_2O$) precipitation, which can be implemented in wastewater treatment plants that use enhanced biological or semi-biological/chemical phosphorus (P) removal. Struvite/MAP is formed by a basic precipitation reaction in different stages of the wastewater treatment process, where magnesium ($Mg^{2+}$), ammonium ($NH_4^+$), and orthophosphate ($PO_4^{-3}$). This chapter aims to discuss: (i) the progress in extraction of P from sewage sludge and animal manure; (ii) the methods to create optimal conditions for struvite precipitation in such media; (iii) the avenues for overcoming the problems associated with choosing the right Mg source, pH adjustment and the non-acceptable level of organic matter in the initial suspension; and (iv) the implication of struvite as an alternative fertilizer for the global agriculture sector.

**Keywords:** phosphorus recovery, phosphate rocks, alternative fertilizers, wastewater, struvite

## 1. Introduction

Phosphorus (P) is an essential nutrient for all forms of life. Specifically, P is an important nutrient element in agriculture and a major limiting factor for plant growth and the entire food production chain. Currently, both P and nitrogen (N) are the basic components of mineral fertilizers viewed as an irreplaceable part of modern agriculture. P is a resource obtained mainly from phosphate rocks located in a few regions of the world [1]. More than 87% of all mined and processed phosphate rock is turned into fertilizers while few phosphate rock is used for additives in livestock feed and food.

There are evidence that P reserves used for mineral phosphate fertilizers as a primary source of P input to agricultural lands are steadily decreasing with time, with the expectation to be depleted in a few centuries or less [2–4]. More striking is the view of the United States Geological Survey according to which the phosphate deposits will last about 50 years at the current rate of extraction [5]. Such a perspective can

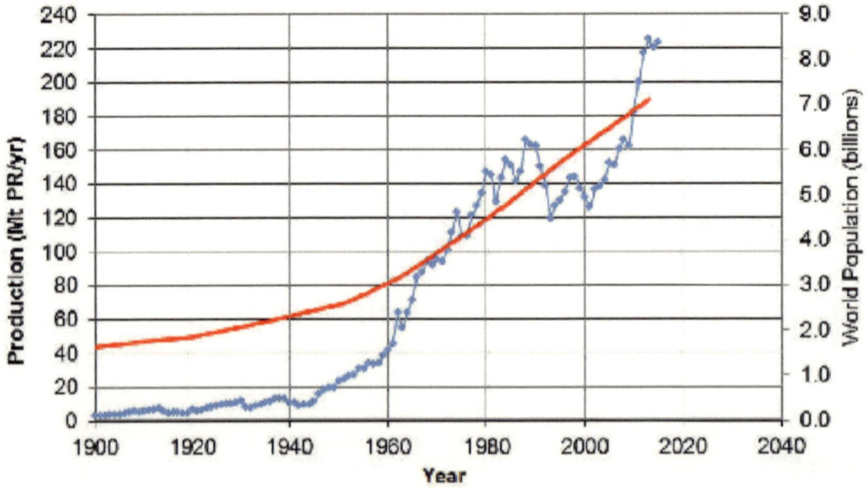


**Figure 1.**
*Global production of phosphate rock (blue) coupled with world population (red) in time [6].*

be reasonably explained by the continuously growing population and rising global demand for food [6]. Global production of phosphate rock coupled increased world populace is shown in **Figure 1**.

The expected P rocks deficiency is supported by the current monetary evaluation of the global phosphate rock market. It shows an annual growth rate higher than 5% in recent years [7] reaching $25.49 billion in 2023.

Recognizing that phosphate rock is a limited resource for mineral fertilizers production, the necessity of identifying P-rich waste streams and finding technical ways for P extraction into valuable products is of extreme importance. Part of the increased pressure on P resources could be alleviated by recycling P contained in various agricultural, industrial, and urban wastes [8, 9]. Excess P during fertilization and the non-proper treatment of wastes, ground an imbalance of the biosphere following severe negative environmental effects. While previous efforts have been directed toward removing P from the wastewater discharged into surface water, current efforts are directed to recover P as useful species [10]. The potential of the waste P is high as the global mass balance shows that up to 20% of the world's P-production (about 3Mt P/year) is currently lost [11]. On the other hand, resource recovery from waste streams is increasingly seen as one option to improve the economics of wastewater treatment. Domestic sewage, industrial wastewater, and manure are no longer considered waste but just a resource of different origins [12]. The green line in **Figure 2** (below) outlines the recovery route in addressing the P-resource recycling in the framework of the circular economy concept.

## 2. Phosphorus recovery as struvite

The shift into circular use of P passes through the implementation of multidimensional innovations focused on the creation of P recovery/recycling technologies [14]. The main well-recognized "waste" streams containing nutrients and causing negative

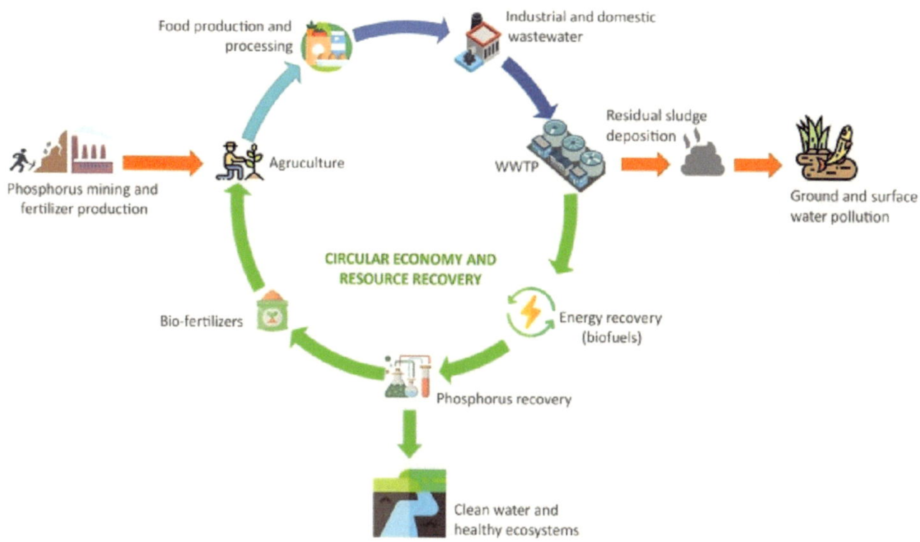


**Figure 2.**
*Phosphorus production, consumption, and recycling pathways [13]. The color code indicates the impact in terms of environmental health and sustainability: Blue – Neutral (currently applied practices); red – Negative practices to be avoided or minimized; green – Processes with positive impact.*

environmental effects include animal manure, urban wastewater and sewage sludge, and food processing wastewater [15]. In the following sections, this chapter discusses two specific potential sources of P, namely the sewage and manure waste streams.

Domestic wastewater and sludge could be regarded as an important secondary source of P. The sludge water from municipal wastewater treatment plants (WWTP) is a fluid containing a reasonable level of P (120–160 mg/l $PO_4^{3-}$), which allows the extraction of this nutrient with good economic effect, especially taking into account the global volumes of this waste stream. According to many authors, its potential counts for 15–20% of the global phosphorus demand [8, 16, 17]. For this reason, phosphate recovery has high priority in sewage treatment. European Union estimations show that the sludge amount produced annually is over 7 Mt. dry solid (DS), while globally, 1.3 Mt. P/year is treated in WWTPs worldwide [8, 17]. Animal manure production is another significant waste stream containing P. The amount of P in manure compared to municipal wastewater depends strongly on the manure source and type. Common *o*-phosphate concentrations of manure liquid fraction varied in a wide range of 50–1200 mg/l. However, the high suspended and dissolved organic matter should be taken into consideration in choosing the P recovery option. The available data show that in evaluating the by-products and waste sources of P in Mt. per year (**Table 1**) P in animal manure is almost one order of magnitude higher in sewage sludge (20–30 Mt. vs. 3–5 Mt., respectively) [19].

| Molar ratio $Mg^{2+}$: $PO_4^{3-}$ | 1:1 | 2:1 | 3:1 | 1:1 | 2:1 | 3:1 | 1:1 | 2:1 | 3:1 |
|---|---|---|---|---|---|---|---|---|---|
| pH | 8 | 8 | 8 | 9 | 9 | 9 | 10 | 10 | 10 |
| P Removal efficiency, % | 45.6 | 44.7 | 44.9 | 76.1 | 76.3 | 78.4 | 79.9 | 89.9 | 89.2 |

**Table 1.**
*Series of experiments using supernatants produced after centrifugation of sludge from WWTP-Burgas [18].*

Interestingly, several technologies for P recovery from wastewater and sludge are operating at either full or demonstration scale. However, P-recovery technologies are focused mainly on the aqueous phase of sludge (the so-called sludge water). In the solid phase, the technologies are directed either to recover P from the dewatered sewage sludge or from mono-incinerated sewage sludge ashes [20]. The accepted means differ in respect of technology choice, costs, efficiency, and product purity.

Chemical precipitation of P into salts of low solubility is a common method for removing dissolved phosphorus from wastewater: in the form of magnesium ammonium phosphate (MAP) hexahydrate ($MgNH_4PO_4{\cdot}6H_2O$), also called struvite, or calcium phosphates [21]. Struvite has the advantage of being a slow-dissolving salt, while calcium phosphates are characterized with extremely low water solubility. P-recovery is usually based on the chemical precipitation of struvite from concentrated wastewater or the liquid fraction remaining after anaerobic digestion of sludge in WWTP. The chemical precipitation of struvite removes up to 98% of the soluble phosphates and 20–30% of the soluble ammonia contained in the targeted liquor [22]. Notably, the main challenge in relation to struvite precipitation is the P-recovery from wastewater characterized by phosphorus concentration of less than 50 mg/L and suspended solids concentration (TSS) above 1000 mg/L [23].

The struvite formation process is a function of pH and the molar ratio between magnesium, ammonium, and phosphate ions. The precipitation occurs in alkaline conditions and optimal crystal formation is observed at pH above 9 and equimolarity of the constituent ions [24]. However, several studies show that struvite can be synthesized in a wide range of pH values. Even experiments applying a pH slightly above 8 show positive results, but the formation and precipitation rates are lower compared to the optimal pH range. This is confirmed by studies with real dewatered sludge liquor (DSL) taken from the MWWTP of Burgas, Bulgaria (initial $PO_4^{3-}$concentration of 86.7 mg/l) [18]. In the same study, $MgCl_2$ was used as a precipitation agent while different pH and molar ratio $Mg{:}PO_4$ were applied (**Table 1**). Obviously, both pH and mole ratio $Mg{:}PO_4$ are crucial factors affecting precipitation efficiency. However, the highest effect is observed at pH near and above 9.0. At extremely high pH values (above 11), the struvite yield decreased due to the formation of $Mg(OH)_2$ and transformation of ammonia ions into free $NH_3$ (reducing the general availability of $Mg^{2+}$ and $NH_4{}^+$ in the medium). Experiments carried out by Saidou et al. [25] showed that at initial solution pH of 10, another phosphate mineral, namely $Mg_3(PO_4)_2.2H_2O$ starts precipitating. Such formation results in other dominant species formation. In fact, varying the pH levels results in different species of phosphate ions. The precipitation of struvite requires equal molar ratios of its components: magnesium, ammonia, and phosphate ions. Different molar ratios have been shown to influence several characteristics of the struvite. It could be argued that as more Mg is available, more crystal units are generated, hence larger crystals could be formed. As reported by Merino-Jimenez et al. [26], several studies have applied the ratios of phosphate and magnesium ions around 1:1.2 for optimum struvite yield. The latter result confirms the effects shown in **Table 1**.

As it was shown above, one of the main factors for struvite formation is the pH. In the targeted P-containing waste streams, approximately 90% of P is trapped in sewage sludge following primary and secondary sedimentation [27]. The pH of these fluids (usually centrate) is in the range of 7.3 and 7.5; far from the optimal pH values providing technologically acceptable conditions for MAP crystals formation. In this case, as a general option struvite crystallization is achieved through alkalization by reagents such as NaOH.

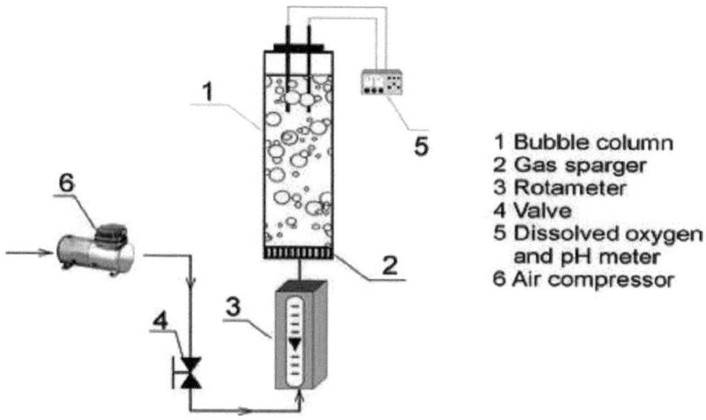


**Figure 3.**
*Air stripping unit design [28].*

Besides the direct reagent alkalization there, pH can be elevated by carbon dioxide ($CO_2$) stripping through aeration of the reagent mixture. **Figure 3** shows the principal scheme of the striping system [28]. The mechanism of this treatment is based on the manipulation of the dissolved carbon dioxide mass balance in the liquid. The aeration of the solution leads to the decomposition of the carbonic acid ($H_2CO_3$) to $H_2O$ and $CO_2$ which results in pH elevation. The described mechanism seems to be appropriate in the case of sludge water treatment as it contains high amounts of organic matter which potentially enriches the liquid with $CO_2$ due to the microbial activity occurring in the suspension. It was found that the rate of pH elevation depends strongly on the spangling area of the air distribution system while the air flow rate does not influence considerably the dissolved oxygen level which governs the $CO_2$ stripping process [28]. The theoretically calculated values of the volumetric mass transfer coefficient have been compared with those obtained experimentally. Based on the data obtained, relationships of pH/$k_La$ (mass transfer coefficient) were developed. These correlations serve as a tool for the prediction of pH during the struvite precipitation process. The pH dependence of air rate is given in **Figure 4**.

The results showed evidently that within the volumetric air rates applied pH elevation from 7.5 to 8.4 was achieved by $CO_2$ stripping in less than 25 minutes. Such a retention time in the reactor is technologically applicable. Actually, a pH value around 8.5 is high enough for effective struvite precipitation even in several studies the optimal pH for the process is shown to be in the range of 9 to 9.5. The targeted pH 9 could be also achieved in an acceptable retention time of less than 1 hour. On the other side, the slow pH change during the aeration could be considered an advantage of the $CO_2$ stripping process because it restricts the rapid increase of solution saturation. At such conditions, the struvite crystallization process predominates, and in addition, Ronteltap et al. [29] reported that the crystal sizes at pH values of 7–11, and found that the largest crystals occurred at pH 8.

The concentration of magnesium is the other limiting factor for struvite crystal formation, the choice of Mg source is important as it forms nearly 75% of the struvite production costs [21]. Several commercial magnesium salts such as $MgSO_4$, $MgCl_2$, MgO, or $Mg(OH)_2$ have been used to precipitate struvite from different liquid wastes [30]. Cheaper options based on seawater as a source of Mg ions are thoroughly studied as well demonstrating a comparatively high phosphorus recovery rate (80–90%) and precipitation of several different products with the domination

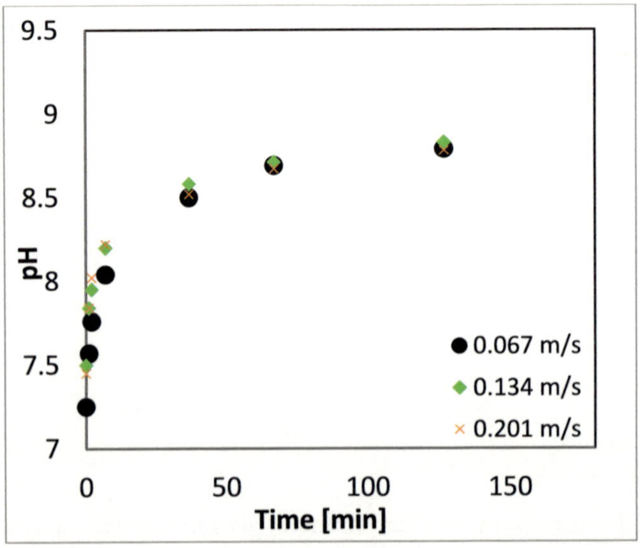


**Figure 4.**
*Elevation in pH by air stripping through applying different volumetric air rates [28].*

of struvite (but also containing magnesium calcite and calcite due to the complex ion composition of the seawater) [31].

Similar results were obtained in the research group of the authors, where seawater brine was used to precipitate P from WWTP sludge centrate [32]. Also, the seawater brine used was found to contain significant amounts of calcium and potassium (**Table 2**). However, the magnesium concentration is higher (59.5 g/l) than the calcium concentration (3.5 g/dm3) which resulted in effective struvite precipitation with minimal calcium phosphate co-production during the process.

The MAP crystals obtained by sweater brine used during this study were compared with struvite produced via the application of a conventional magnesium source ($MgCl_2.6H_2O$). Starting from a model solution of $(NH_4)_2HPO_4$ the precipitation was carried out at a mole ratio Mg:P = 2:1 and pH of 9.5, within 15 minutes of continuous slow mixing (50 rpm). The crystals obtained following a period of 30 minutes for struvite agglomeration, were studied under a scanning electron microscope (SEM) (**Figures 5** and **6**).

The SEM images showed that in both cases the crystals obtained are of typical struvite morphology. Brine-induced crystals have an average size (length) of 280–300 μm, while the crystals precipitated by $MgCl_2$ are of size close to 200 μm. The difference can be explained by the calcites produced during the precipitation of seawater brine.

| Elements | $Na^+$ | $Mg^{2+}$ | $SO4^{2-}$ (as S) | $K^+$ | $Ca^{2+}$ | $Cl^-$ |
|---|---|---|---|---|---|---|
| Concentration ins seawater brine [g/L] | 11.34 | 59.5 | 19.17 | 16.03 | 3.5 | 161.97 |

**Table 2.**
*The concentration of main elements present in the seawater brine [32].*

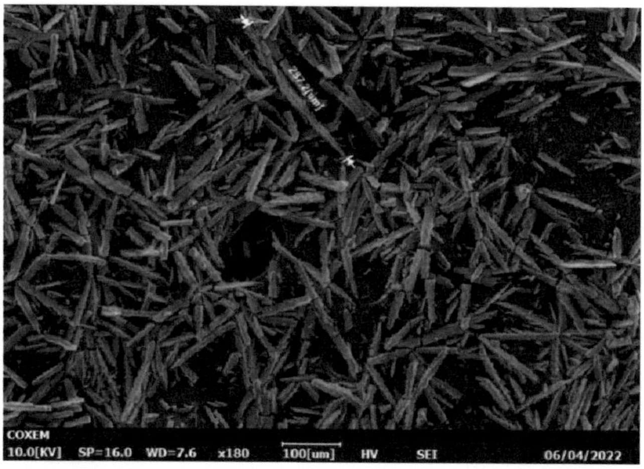


**Figure 5.**
*Struvite crystals precipitated by seawater brine.*

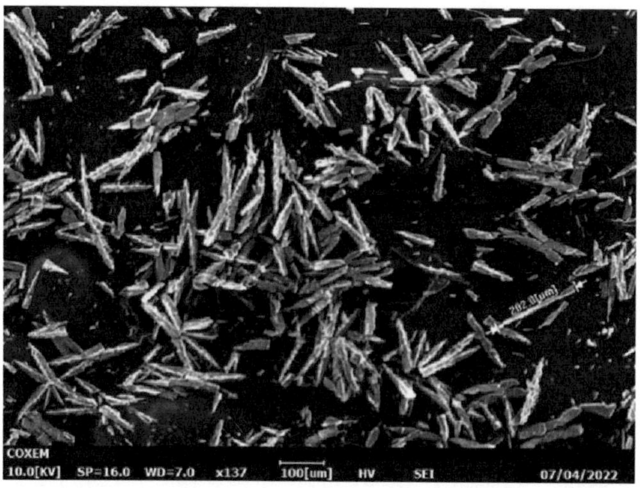


**Figure 6.**
*Struvite crystals precipitated by $MgCl_2.6H_2O$.*

## 3. Influence of organic matter content

The particulate and colloidal organic matter is known to slow the kinetics of struvite formation. Even a low concentration of organic matter has an adverse, inhibiting effect on crystal growth [31, 33–35]. To this end, some promising results for the separation of the nutrients ions from the organic matter dissolved and suspended in the wastewater streams were already obtained in our recent studies [36]. Membrane separation was used as a process to improve wastewater characteristics for further struvite crystallization. The most commonly used membrane processes for organic matter separation are Microfiltration (MF) and Ultrafiltration (UF). It was found that if dead-end MF membranes are used a fast membrane permeability decrease and

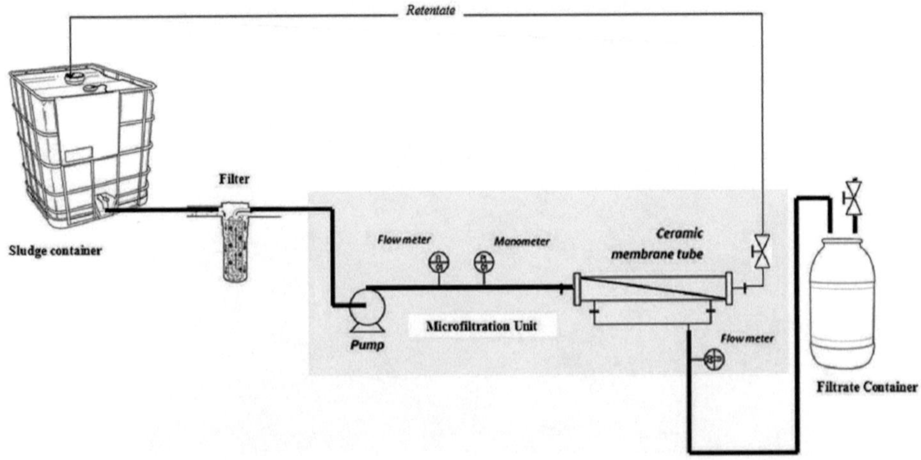


**Figure 7.**
*Membrane filtration unit [32].*

sharp loss of phosphates is observed (case study for filtration of municipal WWTP sludge dewatering liquid and swine). Positive results were obtained by the application of ceramic ultrafiltration membranes (**Figure 7**). The UF unit has been operated in a cross-flow mode at a pressure of up to 6 bars and a recirculating flow rate of 20%. The UF module was loaded with sludge water with the following characteristics: COD – 78.1 $gO_2$/l and TSS - 2.6 g/L; phosphorus and ammonia levels of 250 mg/L and 2.4 g/l, respectively. Two types of ceramic tube membranes, namely with pore sizes of 50 nm and 200 nm, were used [32].

The volumetric rate of the permeate stream did not change during all filtration runs. In the case of usage of a 200 nm UF module, the permeate parameters were: phosphates –195 mg/l, COD - 642 mgO2/L, TSS - 0.15 g/L, and ammonium ions - 224 mg/l. The organic matter and TSS removal rates observed were over 99% while the phosphate reduction was as low as 22% (**Table 3**). In the case of the 50 nm UF module, the removal efficiencies for COD and TSS were again over 99%, however, the phosphates in the filtration product were much lower (only 48% of the initial total amount). In summary, the results obtained show that the 200 nm UF module is more appropriate as it keeps a higher level of phosphorus in permeate, i.e. higher recovery and struvite yield is expected.

| UF membrane (pore size, nm) | Initial COD in the sludge, $mgO_2$/l | Residual COD in permeate, $mgO_2$/l | COD removal level, % | Concentration of $PO_4$ in permeate, mg/l |
|---|---|---|---|---|
| 50 | 78,100 | 414 | 99.47 | 118 |
| 200 | 78,100 | 642 | 99.18 | 195 |

**Table 3.**
*Results obtained through the application of the nano-filtration process.*

## 4. Case studies of struvite production

The most common methods for P-removal from domestic wastewater treatment are Chemical Phosphorus Removal (CPR) and enhanced biological P removal (EBPR) [37]. Currently, the biological process for phosphate removal is more attractive as the rejected sludge liquor after digestion units have a higher phosphorus content with higher phosphorus recovery. In addition, the CPR technique has the disadvantage of an increased sludge volume (by 26% compared to EBPR) with reduced dissolved P content. However, in several cases, the less complicated approach, the CPR, is still the preferred approach, mainly due to its high efficiency in achieving the requirements for residual P in the WWTP outlet streams. Contrary, even taking into account its higher sustainability, the EBPR can rarely provide P removal efficiency higher than 60%. It is well recognized that the combination of EBPR and anaerobic sludge digestion offers a better opportunity for P extraction because, during the anaerobic stage, P is released to a high extent and makes the following struvite precipitation more efficient. The eventual struvite precipitation after the combination of chemical and biological phosphorus removal is shown herein by two case studies aiming at the WWTP of Burgas City in Bulgaria (applying both chemical and biological P removal) and the municipal WWTP in the town of Pomorie. Bulgaria which applies solely biological process for P removal. Both plants are serving settlements located on the South-East Coast of the Black Sea. The process used for both is a conventional activated sludge system with denitrification/nitrification zones (**Figure 8**). The conditions in this system configuration allow partial bio-dephosphatation. However, in the case of Burgas WWTP, $FeCl_3$ is added before the activated sludge basin (doses ranging from 1.7 to 2.4 mg$Fe^{3+}$/mgP) for the chemical precipitation of phosphates. This method has one additional drawback i.e. locking the phosphates into the insoluble form of $FePO_4$, which needs to be treated for re-mobilization of the phosphate ions.

The data shown in **Table 4** reveals that a considerable part of phosphates remain in the returned streams and are available to be recovered.

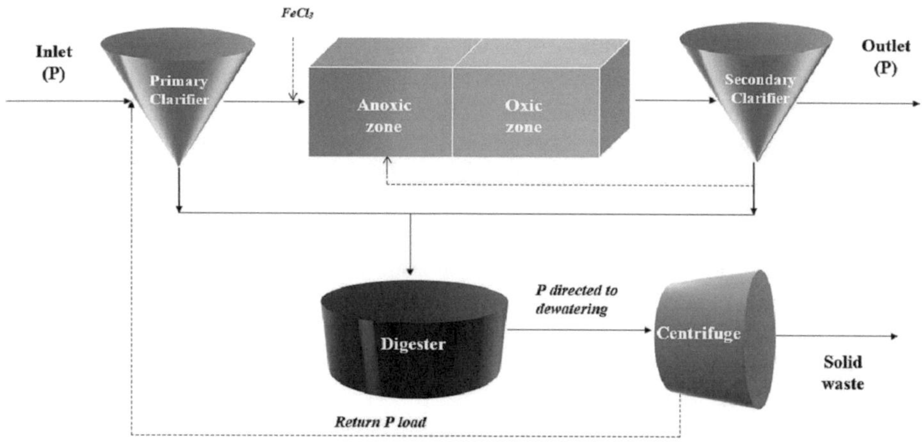


**Figure 8.**
*Principle scheme of the WWTPs under study.*

| WWTP Name | Inlet P [kg P/d] | Outlet P [kg P/d] | P directed to dewatering system [kg P/d] | Solid waste [kg P/d] | Returned P load [kg P/d] |
|---|---|---|---|---|---|
| Burgas | 103 | 23 | 120 | 80 | 40 |
| Pomorie | 29 | 15 | 26 | 14 | 12 |

**Table 4.**
*Phosphorus balance.*

| WWTP | Phosphate ions, mg/l | Ammonium ions, mg/l | Expected struvite production from*, kg per day sewage waste |
|---|---|---|---|
| Pomorie | 138 ÷ 250 | 250 ÷ 600 | 104 |
| Burgas | 86 ÷ 131 | 250 ÷ 500 | 368 |

**If the Average dewatering liquor P concentration for WWTR Pomorie and Burgas are 200 mg/l and 100 mg/l respectively.*

**Table 5.**
*Phosphate and ammonia levels in the centrate.*

Based on the average phosphate and ammonia levels in the centrate, the amount of struvite produced in the corresponding plants has been calculated (**Table 5**). The struvite production estimation is based on experimental results obtained for optimal Mg/P molar ratios and pH 9 when seawater brine is used as a source of magnesium ions (according to the characteristics mentioned above in the text).

## 5. Struvite as fertilizer

One of the main concerns of applying alternative fertilizer obtained by P-recovery from waste streams is related to the potential presence of pathogens, heavy metals, and also low levels of bio-utilization [38]. However, these drawbacks are case-specific and could be solved on a technological level. Usually, the heavy metal content of struvite is below the detection limit and it is suggested that is mainly linked to total organic matter in the samples and not to the struvite crystals themselves [39]. It is already known that MAP is suitable for feeding decorative plants like grass, tree seedlings, decorative plants, vegetables, flowers, and grass gardens [40]. The main question for the agriculture industry remains the efficiency of alternative fertilizers, such as struvite, when they are used to replace (or complement) commercially available products.

The chemical nature of struvite defines it as a combined fertilizer source of nitrogen, phosphorus, and magnesium which is an important microelement for the photosynthetic systems of plants [41]. Due to the limited water solubility ($K_{sp}$ value of 7.59 × 10–14), it is also a natural slow-releasing fertilizer that could provide a steady supply of plant nutrients over an extended period of time. The nutrient release could be significantly influenced and regulated by the soil microbiology and physical characteristics matching the plant utilization rate [42]. Contrarily, conventional mineral fertilizers are readily soluble and could be washed out by rainfalls and irrigation increasing the risk of surface and groundwater pollution and eutrophication [43].

Recently, the author's research group tested struvite fertilization toward maize growth which was followed for 6 months of vegetation in a test field [44]. The quantity and quality of the crops harvested were compared with control samples

| Fertilizer | Yield, kg/ha | $M_{1000}$, g |
|---|---|---|
| Control sample | 51.43 | 238.92 |
| Ammonium nitrate | 55.87 | 246.40 |
| Carbamide + Ammonium nitrate | 56.64 | 249.20 |
| Struvite | 54.60 | 245.27 |
| Struvite + Ammonium nitrate | 55.11 | 246.36 |

**Table 6.**
*Productivity of maize applying different fertilizers ($M_{1000}$ – The absolute weight of 1000 grains).*

| Fertilizer | Wet, % | Proteins, % | Fats, % | Starch, % |
|---|---|---|---|---|
| Control sample | 14.1 | 8.8 | 3.4 | 58.9 |
| Ammonium nitrate | 14.2 | 9.5 | 3.3 | 59.0 |
| Carbamide + Ammonium nitrate | 14.1 | 9.5 | 3.2 | 58.5 |
| Struvite | 14.0 | 8.8 | 3.2 | 58.2 |
| Struvite + ammonium nitrate | 14.1 | 9.7 | 3.2 | 57.5 |

**Table 7.**
*Qualitative characteristics of maize while using different fertilization options.*

obtained by cultivation with conventional fertilizers. The evaluation was performed based on the yield and nutritional characteristics of the corn. According to the results obtained, the struvite is a very efficient complement to the nitrogen fertilizers (such as ammonium nitrate and carbamide) as an alternative to the normal superphosphate and triple superphosphate in the role of P source. Considering overall P availability, different struvite samples show patterns with a continuous soil release of phosphorus without peak concentrations followed by fast decline which is typical for super phosphates or rock phosphate for instance [45] (**Tables 6** and 7).

## Acknowledgements

The authors wish to heartily acknowledge the support under the BG05M2OP001-1.002-0019 *Clean technologies for sustainable environment – waters, waste, energy for the circular economy* "*(Clean&Circle)* Project financed by the Operational program "Science and Education for Smart Growth" 2014-2020, co-financed by the European Union through the European structural and investment funds.

## Author details

Valentin Nenov*, Hyusein Yemendzhiev and Gergana Peeva
Department of Chemical Technologies, Burgas "Asen Zlatarov" University, Burgas, Bulgaria

*Address all correspondence to: vnenov@btu.bg

## References

[1] Wzorek Z. Alternative phosphorus raw materials. Przemysł Chemiczny. 2006;**85**:880-882

[2] Smit AL, Van Middelkoop JC, Van Dijk W, Van Reuler H. A substance flow analysis of phosphorus in the food production, processing and consumption system of the Netherlands. Nutrient Cycling in Agroecosystems. 2015;**103**:1-13

[3] Cordell D, White S. Tracking phosphorus security: Indicators of phosphorus vulnerability in the global food system. Food Security. 2015;7:337-350

[4] Sverdrup HU, Ragnarsdottir KV. Challenging the planetary boundaries. II. Assessing the sustainable global population and phosphate supply, using a system dynamics assessment model. Applied Geochemistry. 2011;**26**:307-310

[5] Phosphate Rock Statistics and Information. 2023. Available from: https://www.usgs.gov/centers/national-minerals-information-center/phosphate-rock-statistics-and-information

[6] Carvalho FP. Pesticides, environment, and food safety. Food and Energy Security. 2017;**6**(2):48-60

[7] de Ridder M, de Jong S, Polchar J, Lingemann S. Risks and Opportunities in the Global Phosphate Rock Market. Netherlands: The Hague Centre for Strategic Studies (HCSS); Rapport No17 | 12 | 12 ISBN/EAN: 978-94-91040-69-6

[8] European Commission. EU Consultative Communication on the Sustainable Use of Phosphorus. European Commission; 2013. pp. 1-19. Available from: https://ec.europa.eu/environment/consultations/pdf/phosphorus/EN.pdf

[9] Van Dijk KC, Lesschen JP, Oenema O. Phosphorus flows and balances of the European Union memberstates. Science of the Total Environment. 2016;**542**:1078-1093

[10] Cichya B, Kuźdżał E, Krztoń H. Phosphorus recovery from acidic wastewater by hydroxyapatite precipitation. Journal of Environmental Management. 2019;**232**:421-427

[11] Alewell C, Ringeval B, Ballabio C, et al. Global phosphorus shortage will be aggravated by soil erosion. Nature Communications. 2020;**11**:4546. DOI: 10.1038/s41467-020-18326-7

[12] European Investment Bank - Environment and Natural Resources Department. Wastewater as a resource 2022. Available from: https://www.eib.org/attachments/publications/wastewater_as_a_resource_en.pdf

[13] Nenov V, Peeva G, Yemendzhiev H, Stancheva M, Zerrouq F. Phosphorus consumption. From linear to circular flow. Moroccan Journal of Chemistry. 2020;**8**(4):819-829

[14] Smits MJW, Woltjer GB, Luesink HH. Phosphorus Recycling from Manure – A Case Study on the Circular Economy: Work Package 4. EU; 2018. 61 p. Available from: https://circular-impacts.eu/sites/default/files/D4.5_Case-Study-Nutrient-Recycling_FINAL.pdf

[15] Marin E, Rusănescu CO. Agricultural use of urban sewage sludge from the Wastewater Station in the municipality of Alexandria in Romania. Water. 2023;**15**(3):458. DOI: 10.3390/w15030458

[16] Yuan Z, Pratt S, Batstone DJ. Phosphorus recovery from wastewater through microbial processes.

Current Opinion in Biotechnology. 2012;**23**(6):878-883

[17] Cassardo C, King A. Global warming and water sustainability. E3S Web of Conferences. 2014;**2**:02006

[18] Nenov V, Jemendjiev H, Peeva G, Bonev B, Zerouq F. Struvite recovery options in conventional wastewater treatment plants. Journal of Materials and Environmental Science. 2016;7(1):113-122

[19] Oenema O, Chardon W, Ehlert P, Van Dijk K, Schoumans O, Rulkens W. Phosphorus fertilizers from by-products and wastes. The International Fertiliser Society at a Meeting in Cambridge. 2015;**717**:1-56

[20] Luyckx L, Van Caneghem J. Recovery of phosphorus from sewage sludge ash: Influence of incineration temperature on ash mineralogy and related phosphorus and heavy metal extraction. Journal of Environmental Chemical Engineering. 2012;**9**(6):106471

[21] Zhou K, Barjenbruch M, Kabbe C, Inial G, Remy C. Phosphorus recovery from municipal and fertilizer wastewater: China's potential and perspective. Journal of Environmental Science. 2017;**52**:151-1597

[22] Kabdaşli I, Safak A, Tünay O. Bench-scale evaluation of treatment schemes incorporating struvite precipitation for young landfill leachate. Waste Management. 2008;**28**(11):2386-2392. DOI: 10.1016/j.wasman.2007.10.020

[23] Li M, Zhang H, Sun H, Mohammed A, Liu Y, Lu Q. Effect of phosphate and ammonium concentrations, total suspended solids and alkalinity on lignin-induced struvite precipitation. Scientific Reports. 2022;**12**(1):2901. DOI: 10.1038/s41598-022-06930-0

[24] Miles A, Ellis TG. Struvite precipitation potential for nutrient recovery from anaerobically treated wastes. Water Science and Technology. 2001;**43**:259-266

[25] Saidou H, Moussa SB, Amor MB. Influence of airflow rate and substrate nature on heterogeneous struvite precipitation. Environmental Technology. 2009;**30**:75-83

[26] Merino-Jimenez I, Celorrio V, Fermin DJ, Greenman J, Ieropoulos I. Enhanced MFC power production and struvite recovery by the addition of sea salts to urine. Water Research. 2017;**109**:46-53

[27] Biöcher C, Niewersch C, Melin T. Phosphorus recovery from sewage sludge with a hybrid process of low pressure wet oxidation and nanofiltration. Water Research. 2012;**46**(6):2009-2019

[28] Radev D, Peeva G, Nenov V. pH control during the Struvite precipitation process of wastewaters. Journal of Water Resource and Protection. 2015;7(16):1399-1408

[29] Ronteltap M, Maurer M, Hausherr R, Gujer W. Struvite precipitation from urine–influencing factors on particle size. Water Research. 2010;**44**(6):2038-2046

[30] Liu B, Giannis A, Zhang J, Chang V, Wang JY. Characterization of induced struvite formation from source-separated urine using seawater and brine as magnesium sources. Chemosphere. 2013;**93**:2738-2747

[31] Shaddel S, Grini T, Ucar S, Azrague K, Andreassen JP, Østerhus SW. Struvite crystallization by using raw seawater: Improving economics and environmental footprint while maintaining phosphorus recovery and product quality. Water Research. 2020;**173**:115572. DOI: 10.1016/j.watres.2020.115572

[32] Peeva G, Yemendzhiev H, Nenov V. Precipitation of struvite from digested sewage sludge by sea water brine and natural flocculants. Journal of Chemical Technology and Metallurgy. 2022;**57**(2):310-318

[33] Hövelmann J, Putnis CV. In situ nanoscale imaging of struvite formation during the dissolution of natural brucite: Implications for phosphorus recovery from wastewaters. Environmental Science and Technology. 2016;**50**:13032-13041

[34] Capdevielle A, Sýkorová E, Béline F, Daumer ML. Effects of organic matter on crystallization of struvite in biologically treated swine wastewater. Environmental Technology. 2016;**37**(7):880-892. DOI: 10.1080/09593330.2015.1088580

[35] Chen RF, Liu T, Rong HW, Zhong HT, Wei CH. Effect of organic substances on nutrients recovery by Struvite electrochemical precipitation from synthetic anaerobically treated swine wastewater. Membranes (Basel). 2021;**11**(8):594. DOI: 10.3390/membranes11080594

[36] Stancheva M, Peeva G, Nenov V, Yemendzhiev H. Membrane filtration pretreatment preceding struvite precipitation. Moroccan Journal of Chemistry. 2020;**8**(3):658-664

[37] Peeva G, Koleva R, Stancheva M, Yemendzhiev H, Nenov V. Phosphorus recovery from waste as struvite. Ecological Engineering and Environment Protection. 2020;**2**:34-44

[38] Vogel C, Rivard C, Wilken V, Muskolus A, Adam C. Performance of secondary P-fertilizers in pot experiments analyzed by phosphorus X-ray absorption near-edge structure (XANES) spectroscopy. Ambio. 2018;**47**(1):62-72. DOI: 10.1007/s13280-017-0973-z

[39] Huygens D, Saveyn H, Eder P, Sancho LD. Wageningen University & Research – Wageningen Environmental Research Technical Proposal [Internet]. Available form https://edepot.wur.nl/440985

[40] Ponce G, Garcia-Lopez R. Evaluation of Struvite as a fertilizer: A comparison with traditional P sources. Agrochimica. 2007;**51**:301-308

[41] Zeiger E, Taiz L. Ch. 7: Topic 7.11: Chlorophyll biosynthesis. In: Plant Physiology. 4th ed. Sunderland, MA: Sinauer Associates; 2006 ISBN 0-87893-856-7

[42] Talboys PJ, Heppell J, Roose T, Healey J, Jones D, Withers P. Struvite: A slow-release fertiliser for sustainable phosphorus management. Plant and Soil. 2016;**401**:109-123

[43] Monteagudo L, Moreno JL, Picazo F. River eutrophication: Irrigated vs. non-irrigated agriculture through different spatial scales. Water Research. 2012;**46**(8):2759-2771. DOI: 10.1016/j.watres.2012.02.035

[44] Peeva G, Yemendzhiev H, Koleva R, Nenov V. Agrotechnical assessment of Struvite application. Journal of Agricultural Chemistry and Environment. 2021;**10**:213-221

[45] Muys M, Phukan R, Brader G, et al. A systematic comparison of commercially produced struvite: Quantities, qualities and soil-maize phosphorus availability. Science of theTotal Environment. 2021;**756**:143726. DOI: 10.1016/j.scitotenv.2020.143726